The Politics of Fishing in Britain and France

MICHAEL SHACKLETON
Administrator in the Secretariat of the European Parliament in Luxembourg

Gower

Published by
Gower Publishing Company Limited
Gower House
Croft Road
Aldershot
Hants GU11 3HR
England

Gower Publishing Company
Old Post Road
Brookfield
Vermont 05036
USA

British Library Cataloguing in Publication Data

Shackleton, Michael
The politics of fishing in Britain and France.
1. Fishery policy——Great Britain
2. Fishery policy——France
I. Title
338.3'727'0941 SH255

ISBN 0-566-05161-3

Printed in Great Britain by Blackmore Press, Shaftesbury, Dorset

THE POLITICS OF FISHING IN BRITAIN AND FRANCE

Contents

Chapter Nine - State and society in the fishing industry: an overview

Tables and figures

Acknowledgements and declaration

ACKNOWLEDGEMENTS

The author would like to thank everybody who helped to make the preparation of this study possible: the interviewees who were so generous with their time and information (cf Appendix III); the staff at Le Marin and the CCPM in Paris and at Fishing News in London who always found me a chair and a table to work on; and all those who typed successive drafts over the years. Pat Cooke at the Open University brought to fruition in 1984 what Debbie Curtis had begun in 1979 and both cheerfully deciphered a minute handwriting.

The contents of the study have been influenced by many people, especially those who have been willing to read parts of it. Remy Debeauvais, John Eisenhammer, Michael Palmer, Ian Scott and David Steel all gave generously of their time and helped to identify muddled thinking. However, the overall shape of the work owes most to my supervisor Barry Buzan. He never failed to identify weaknesses in style and argument or to point the way towards more fruitful lines of enquiry. Without him a lengthy absence from Britain would surely have proved fatal to the enterprise. The faults that undoubtedly remain are, as ever, the responsibility of the author and the author alone.

DECLARATION

Some of the ideas contained in this study have appeared in the following two articles by the author:

SHACKLETON, M (1981) 'What made the French fishermen resort to open protest' in Marine Policy, Vol.5, No.4, pp 340-343.

SHACKLETON, M.(1983) 'Fishing for a policy? The common fisheries policy of the Community in WALLACE, H. WALLACE, W. and WEBB, C. (eds) Policy Making in the European Community, Chichester, John Wiley, pp 349-371.

All the views expressed in this book are strictly personal and in no way represent the opinion of the European Parliament.

Summary

The last decade has seen dramatic changes in the environment facing the fishing industries of Western Europe. This study looks at Britain and France and compares the response of government and industry to those changes between 1975 and 1983. It argues in the opening chapter that that response can be best understood in terms of the nature of the general relations which link state and society in the two countries. Thus France can be characterised as a state-led society, which has generated protectionist forms of economic policy and a 'dirigiste' style of policy making, where the institutions of the state seek actively to determine the way in which an economic sector develops. By contrast, Britain can be seen as a society-led state, in which a liberal conception of economic policy has been matched by a more consensual style of policy-making, where the agents within a sector are left to develop their own individual responses to change. Chapters two to four consider in turn the impact of political and economic change upon the structure of the two industries, the transformation of the international framework of negotiation within which the two governments dealt with the issue and the development of the institutional links between government and the fishing interest. The chapters that follow (five to eight) are organised around four perspectives on the relations between an interest group and government. These are entitled interventionist, mediatory, direct-action and self-help and each stresses a different aspect of the behaviour of state institutions and a societal interest. In all four chapters, the available evidence is assessed in terms of what we might expect that behaviour to be, given the extent of the change that overtook the industry and the political and economic character of the two countries. The final chapter reviews the distinction beween a state-led society and a society-led state and suggests two conclusions: firstly, that the pattern of relations between industry and government retained its distinctive shape in the two countries, despite severe pressures and secondly, that any judgement of the relative success of the two states and their respective industries in developing a response to change depends on one's appreciation of the merits of two contrasting political and economic philosophies.

1 The character of the study

Introduction

Throughout Western Europe the 1970s were characterized by industrial decline. The international economic environment ceased to be as favourable as it had been in the previous two decades. Many industries were faced with new competitors, who were better able to adapt to change in a period when the overall level of international economic activity was in decline. The effects within Western European society were enormous, not least in the relationship between the industries concerned and their respective governments. Governments tended to alter the emphasis of policy from a macro- to a micro- level. Instead of gearing policy to create the general conditions necessary for the success of particular industries, they found themselves drawn deeper into the problems of individual cases. The industries, for their part, sought to redefine their relationship with government in order to minimise the impact upon them of changes in the international environment. Not surprisingly this changing relationship was not free from mutual recrimination. Those in the industry would complain that government had not done enough to protect them: the cry 'sell-out' was often in the air. Those in government would point to their own lack of room for manoeuvre: for them it was important to display 'realism' and a recognition of the limits of action.

Nowhere was the impact of change more keenly felt than within the Western European fishing sector. It was confronted in the

1970s with an environment which threatened it in three distinct ways. In the first place, there was a growing awareness of resource scarcity, that fish stocks were in serious decline. This was something which had been conspicuously missing hitherto. At the beginning of the decade most people within the industry still shared T.H. Huxley's view expressed at the end of the last century that:

> "... it may be affirmed with confidence that in relation to our present modes of fishing a number of the most important sea fisheries, such as the cod fishery, the herring fishery and the mackerel fishery, are inexhaustible; nothing we could do seriously affects the number of fish."[1]

However, it became increasingly clear that such optimism was ill-founded, as the total world catch levelled out and even slightly declined during the 1970s. Dramatic improvements in technological skills - sonar techniques to locate shoals, more sophisticated gear to catch them and better-equipped ships to process them - all placed greater pressure on the resource and underlined that though renewable, it was not inexhaustible.

The second problem facing the fishing sector was that of the changing rules governing ownership of marine resources. Traditionally it had been supposed that beyond a narrow coastal belt, the sea and the resources within it were available for use by all. Thus no-one could be legitimately prevented from fishing on the open sea wherever, as often and as much as they wished. However, this idea had never been universally accepted and challenges to it grew in the period following World War II. As early as 1952, Chile, Ecuador and Peru issued the Santiago Declaration by which they asserted sovereignty over all activities within 200 miles of their coasts, including fishing. Subsequently, growing awareness of the potential wealth of the oceans coincided with a pattern of encroachment upon the oceans by coastal states.[2] The average size of territorial waters, for example, increased markedly. In 1960, forty-three states still had territorial waters of only three miles; by 1972, the number had declined by eighteen to twenty-five. In the same period the number of states with territorial waters of twelve miles and over, more than doubled from twenty-six to fifty-nine.[3]

The changes in the rules governing the freedom of the seas assumed a more menacing form for the Western European fishing industry during the course of the Third Law of the Sea Conference (UNCLOS III). This conference which opened in New York in 1973, took up the idea of a 200 mile Exclusive Economic Zone (EEZ) around coastal states, within which they would have the right to determine the conditions of access to any economic resources, whether on or under the sea bed or in the water-column. Thus without exercising all the rights associated with territorial waters - "innocent" navigators, for example, could not be interfered with - coastal states could, within the EEZ, effectively claim ownership of all resources, including fisheries, and thereby deny access to outsiders, if they so wished. Though the UNCLOS III Convention, opened for signature in May 1982, has not yet come into force, the concept of an EEZ is now effectively applied as part of the changed pattern of international rules on the sea. As a result 35% of the sea area of the world has come under coastal state authority.[4]

The third and final problem which beset the Western European fishing industries, was the changed economic conditions under which fishing had to be pursued. The fishing sector throughout the world was faced with the fourfold increase in the price of fuel oil in 1973/1974, an event with damaging implications for an industry where the cost of fuel is such a large element in overheads. However, the effects varied considerably as the rules of ownership changed. For some states, the departure of foreign ships from their own EEZs meant reduced pressure on the stock, less difficulty in making large catches and a reduction in fuel consumption; for others, including the British, French and Germans, exclusion from the EEZ of third states meant either searching for alternatives further away or increased effort in their own, already heavily fished, waters. Either way fuel costs would rise: in the first case because of the distance, in the second case because of the increased pressure on the stocks and the consequent need to be away from port longer to make the same level of catch. Moreover, the Europeans suffered in a further way because the reduced costs of states like Canada and Iceland enabled them to export fishery products at prices with which the European fishermen could not compete.

Changes in these three areas - resource availability, resource ownership and economic conditions - affected all those concerned with the fisheries sector but not always in the same way.[5] Exclusion from the grounds of third states made its most dramatic impact on Hull. It had been the most important port in Western Europe in 1970, but by 1980 the vessel owners' association was disbanded and landings reduced to a trickle. For other ports, the problem was one of exposure to international competition and sudden decline in the value of catches. Peterhead, for example, had seen the value of the catch landed increase over three times between 1976 and 1978 but in 1980 and 1981, the effect of imports on prices made it unprofitable for a time for the fishermen even to set sail. Similar contrasts were seen elsewhere in Europe. In France, for example, ports such as Fécamp and La Rochelle, heavily dependent on fishing in the distant waters of states which claimed 200 mile EEZs, witnessed a decline in activity which was not seen in places such as Boulogne. There the level of landings remained fairly steady throughout the 1970s and early 1980s. However, increasing costs, particularly of fuel, created different but still very sensitive problems for Boulogne. Thus in the summer of 1980 it was the centre of the largest protest action by fishermen in the whole of Western Europe.

These various difficulties generated a new, higher level of pressure from the industry upon the governments concerned. The fishermen expected increased help to deal with the changed environment. Ministers, for their part, not only had to balance the claims of other domestic industries in trouble against those of the fishermen but also to acknowledge that they were not free to devise national solutions on their own. The member states of the EEC devoted considerable effort to devising a common policy which would satisfy all concerned and not give an undue advantage to any particular industry. This proved an immensely difficult task: all EEC fishermen felt that they were being disadvantaged to the benefit of their competitors, each industry saw the others as receiving higher levels of subsidy than itself. As a result only in January 1983 was it possible for the ten member states to agree on the overall shape of the Common Fisheries Policy (CFP), some 13 years after the first attempt to establish a Community policy among the six founder members. The sheer length of the negotiations underlines how much more of a political issue fishing had become than it had been before the changes in the wider international environment had occurred.[6]

The interest of this study lies in one particular aspect of this process of politicization, namely, the character of the relationship between fishing industry and government at the national level. That relationship has a double importance. At a theoretical level, the analyst has the opportunity to consider whether the nature of the relationship developed in the way we might expect in a period of such rapid change. In more practical terms, one is obliged to consider whether events could have turned out other than they did, whether government was negligent in its defence of the industry or whether the industry missed opportunities to influence policy, either through inertia or incompetence. To broaden the discussion at both levels, the study adopts an explicitly comparative perspective, concentrating its attention on two states, Britain and France, with fishing industries of roughly similar size and more similar in structure than any other pair of fishing industries in the EEC.

Theoretically, this comparison facilitates the making of distinctions between the way in which an interest and government interrelate; practically, it opens the way to discussion of whether the relationship was better managed and the interests of the industry better protected under one or other form of institutional arrangements. The final chapter of this study will bring these two aspects together by contrasting the notable differences between the two countries in the context of the decline of the two fishing industries. Were they doomed to decline? Were they sold out by their respective governments? Or could they have been better protected if interest and government had managed their relations in a different way?

2. Understanding the response to change

This study assumes that the relationship between the fishing industry and government in the two countries cannot be fully understood by concentrating attention on one or other of the two parties. Thus the discussion will not be cast in terms of a traditional pressure group study, which is most concerned to establish the ways in which an interest attempts to influence government and the variables which determine the success of those attempts.[7] Equally, it will not look primarily at the relationship from the point of view of the government, assuming that the sources of policy are to be found in the debates within and between ministers and ministries.[8] Rather it will be argued that it is necessary to set the relationship in the context of a wider environment which conditions the attitudes and actions of interest and government alike.

The desire to locate the relationship between interests and governments inside a broader framework in this way has generated a considerable body of literature. Two strands within that literature are of particular relevance to this study. The first strand offers the claim that individual relationships between an interest and a government cannot be appreciated without recognising the general level of pressure placed upon government by all interests in society. The democratic basis of Western societies is seen as legitimising increased demands from groups that see themselves to be economically disadvantaged, with the result that governments become 'overloaded' and incapable of responding adequately.[9] A second strand has developed out of the belief that the independence of interest groups is being undermined by the growth of corporatism. Proponents of this view hold that the difficulty of managing the modern economy has encouraged governments to incorporate interests more tightly within the decision-making framework, in order that those affected by decisions can share responsibility for them.[10]

These two sets of arguments share the advantage that they offer a general context within which to discuss a particular case study. However, both make - whether openly or not - the assumption that all Western European states are subject to the same processes and pressures, and that any differences between them are relatively unimportant. Here it will be argued that those differences are significant and that the contrast between Britain and France within the fisheries sector generates evidence that the broader theories need to accommodate. The 'overload' thesis cannot ignore that the extra demands made upon French governments by the fishing industry fitted into a very different pattern of state intervention than that in Britain and that the levels of tolerance to such demands varied greatly as a result. Equally, no discussion of corporatism in this sector can avoid explaining the very different institutional links between industry and government in the two countries and the incorporation of the French fishing industry by the state into a structure of consultation as long ago as 1945. The intention here is not to resolve these difficulties but to generate material in the context of a case study which will help to broaden the debate on general theories about the changing character of policy-making in Western Europe.

In order to be able to raise questions of this kind, this study will argue that developments in the fishing sector need to be set within the broader relationship existing between state and society. This relationship is presented as both similar and different in the two countries. It is similar to the extent that it enjoys a degree of continuity which is not altered by changes of government and minister. In Britain, between 1975 and 1983, a Conservative government succeeded a Labour one and three different ministers (Peart, Silkin and Walker) were responsible for fishing, while in France Giscard d'Estaing was succeeded as President by Mitterand and the fishing portfolio was held by four men (Cavaillé, Le Theule, Hoeffel and Le Pensec). However, these changes did not, it will be claimed, fundamentally change the relationship between state and society. The difference in the relationship is to be seen in its divergent content in the two countries. The state as an idea and a set of institutions plays so much more important a role in France that it can be fairly characterized as a state-led society; by contrast, the weakness of the state in Britain justifies the use of the term society-led state.[11]

To justify these claims, it is necessary to draw attention to the contrast between the position of the state in the Anglo-Saxon (ie British and American) tradition and in the Continental tradition. In France, as in other continental countries, "the state is a central term of political discourse"; in Britain, however well-developed the institutions of the state, the idea of the state plays a much more limited role in politics.[12] Thus Barker has argued that:

> "the state as such does not act in England; a multitude of individuals each separately and severally act ... There is a bundle of individual officials, each exercising a measure of authority under the cognizance of the Courts but none of them, not even the Prime Minister, wielding the authority of the state."[3]

This difference is not simply a question of the strength of ideas. Those ideas underpin the actions of people both inside and outside government. In the case of Britain and France, they have been mediated by historical developments in such a way as to create divergent policy styles and economic orientations. In the French context, the strength of the idea of the state is reflected in an assertive style of policy-making linked to a broadly protectionist outlook towards economic problems. In Britain, by contrast, a more consensual policy style combines with a more liberal economic outlook to underline the weakness of the idea of the state.

The contrast in policy styles can be traced to the very different historical relationship between nation and state in the two countries. Whereas Britain was a single, relatively homogeneous nation before it developed the institutional machinery of a centralised state, France required the constraints of the Napoleonic state to enforce a degree of unity on a heterogeneous nation. This difference has continued to make itself felt. In France, the thrust for unity has been motivated by a recognition of the fragile nature of existing political cohesion. Consequently, policy makers are peculiarly conscious of the need to buttress the authority of the central state.[14] In Britain, both the success of a process of nation building which is much older than the state bureaucracy, and the tradition of a representative Parliament for the transmission of grievances have obviated the need for, and the possibility of, state institutions exercising an equivalent level of authority.

Thus, within a society like France there is a more widespread sense of the legitimacy of public action and those who are in positions of authority consider themselves to be entrusted with the task of identifying, pursuing and proselytizing the values incorporated in the principles of public law. In Britain, by contrast,

> "it has proved impossible to develop a notion of the inherent responsibilities of the official as the embodiment of the state; to emphasize the personal responsibility of the official in law, or to locate personal responsibility, in the managerial sense of accountability for results, in an orderly manner within the administration."[15]

Very different styles of policy making have emerged as a result. The British administrative elite has always been very reluctant to give any kind of leadership or to develop policies on its own. Independent advisory bodies and committees specifically set up to consider particular problems have been much more important than individual civil servants in initiating change in Britain. Their reports can either be allowed to gather dust or be implemented on the basis of clear political approval legitimated by Parliament.

British officials have tended to be drawn into a style of policy-making, based on the 'logic of negotiation'.[16] Within this logic there are no obvious or overriding long-term objectives. Rather decisions emerge out of a process of mutual adjustment with civil servants, politicians and interest groups involved in bargaining over the direction that policy should take. Politicians may appear to have a blueprint for the future on arrival in office but the incremental nature of policy making in the British context generates a constant gnawing away at the blueprint, orchestrated by a civil service more eager to present options than to choose between them.[17]

In France, by contrast, the widespread acceptance of a zone of authority independent of Parliament, parties and pressure groups, has made it easier for the public official to resist particularist claims and to be both more creative and more authoritarian than his British counterpart in search of the public interest. He is not obliged to register political pressure, nor to accept that the only way to govern is to 'muddle through'. As a French civil servant in the Ministry of Industry put it:

> "First we make out a report or draw up a text, then we pass it around discreetly within the administration. Once everyone concerned within the administration is agreed on the final version then we pass this version around outside the administration, Of course, by then it's a _fait accompli_ and pressure cannot have any effect."[18]

Institutional arrangements have come to underline this notion of the independent zone of authority very clearly. Representative and state institutions are firmly separated in the constitutional arrangements of the Republic: the British idea of the "Government in Parliament" has no equivalent in France. This is not to say that there is no debate as to the relationship between the two sets of institutions, but it is a debate premised on separation. And within the state apparatus, every effort is made to underline the special character of the role of those active within it. As Dyson has put it:

> "... institutional innovation like ENA (Ecole Nationale d'Administration) and the French Planning Commission cannot be defined exhaustively in factual terms; they represent an attempt to proselytize 'modernising' values, a part of a conception of coherent, purposive state action."[19]

Not surprisingly British attempts to copy the French example, such as the Civil Service College, have not enjoyed the same degree of success because of the lack of an equivalent cultural background. Equally, one can understand the basis of the claim that "British civil servants, compared with French, generally lack the confidence to operate an imposition relationship with groups".[20]

At the same time, the activist tradition within a society such as France helps to explain the response of society to officialdom. The idea of the state has served as a pole of repulsion as well as a pole of attraction. On the one hand, the idea of the state has generated a strong sense of public purpose with governmental behaviour perceived as serving to lead society forward; on the other hand, the 'limited authoritarianism' implicit in such an activist state role has tended to provoke potential insurrection against authority.[21] The state appears as an entity which has a dynamic of its own, continually seeking to extend its powers and as a result provoking a resistance, which found its intellectual expression in anarcho-syndicalism.

As Birnbaum points out, the fact that the state has never appeared such a powerful element in political life in Britain has meant that such reactions against the state are absent from her history.[22] Consequently there is a much greater 'elasticity' in the political process but one where proposals for change in the status quo are faced with considerably greater resistance, precisely because of the lack of an activist tradition. In Barry's words

> "The result (intended or actual) of a power-diffusing system (like that in Britain) is to raise a series of obstacles to change in the status quo and collective expenditures, thus raising the price (in terms of bargaining costs) of getting collective action."[23]

However, it is not only the character of the development of political institutions and habits in the two countries that has conditioned the distinction between state and society in Britain and France. Both countries have also been profoundly affected by the way in which their economies have evolved and been directed. Here too there are long historical traditions which have continued to influence the behaviour of governments and interests in society and the balancing of public power and private choices in the economic sphere.

The basic difference between the two countries can be described in terms of the contrast between mercantilism or protectionism and free trade as economic doctrines. The former was originally associated with a trade and balance of payments surplus but its essence is "the priority of national economic and political objectives over considerations of global economic efficiency."[24] The latter, by contrast, is based on the assumption that national wealth depends upon the health of the international economy as a whole rather than a calculation of the trade balance of one's own country. Mercantilists, therefore, tend to view economic policy as a zero-sum game where one nation's gains are another's losses, whereas economic liberals perceive it as variable sum with everybody able to gain thanks to the international division of labour.

These two economic philosophies have always aroused very strong disagreements. For much of the post World War II period, it was widely accepted that the protectionist policies implicit in mercantilism had contributed to the build-up of international tension in the 1930s, and to the subsequent outbreak of open hostilities.[25] Hence the economic order that emerged after 1945 was premised on the need to reduce to a minimum the barriers to trade between countries: it was assumed that a more open international economy would help to ensure a more peaceful world. This view only came to be seriously challenged during the 1970s when Western economies ceased to grow at the high rates of the 1950s and 1960s and when domestic industries were threatened by the free flow of imports which a liberal order had made possible. Hence it became respectable once more to favour protection as a way of guaranteeing the welfare of sectors of the economy exposed to international competition.[26]

Britain and France have always taken opposing stances in this debate between liberalism and protectionism. Whereas Britain has combined free trade principles with a limited conception of intervention in the economy, France has adopted a protectionist stance linked to a highly developed sense of dirigisme or state-direction. The reasons for this are linked to their different economic histories. British espousal of liberal doctrine is attributable to the fact that for so long she enjoyed a position of hegemony in the world economy, while France as an industrial late-comer was obliged to adopt a mercantilist approach to catch up.[27] However, the purpose here is to observe the consequences rather than the causes of the difference. Those consequences have been very marked.

Although Britain has consistently favoured the promotion of international trade and commerce, the state itself never assumed a prominent role in that promotion. Rather commercial and trading activities were organized and financed by non-state institutions, notably those in the City of London. It is true that in the more recent past, the general level of intervention by state institutions has developed considerably and no government has sought to argue that the economic health of the country could be assured without any action on its part. Nevertheless, the very fact that the idea of governmental intervention in the economy has been so regularly at the centre of political debate underlines the strength of opposition to the extension of such intervention. As Dyson has put it:

> "Considerable government intervention combines with a hesitancy about the use of its authority; a sense of the 'necessity' of collective action is accompanied by scepticism about its desirability or utility."[28]

The French tradition is completely different. Only in the very recent past has the country sought to expose itself to foreign competition and the market values that it implies. Until then, as Hayward has put it,

> "French governments had for centuries alternated between policies of passive protection and active promotion - state-sponsored captalism and state capitalism - based upon close collusion between the private sector and its public senior partner."[29]

This tradition stretching back to Colbert (whose influence in the fisheries sector will be discussed in Chapter 4) generated a very distinct set of attitudes in the relations between state and society. Whereas the private sector has always looked instinctively for support to the state, no-one has seriously challenged the legitimacy of state action in shaping that support. It is thus not just a question of measuring amounts of money that pass from government to industry but also of observing the way in which that money is perceived by both donor and recipient. Indeed it is for this reason that many observers have argued that despite the change in rhetoric of the Barre government in France between 1976 and 1981 towards a more liberal-sounding conception of economic policy, lip-service only was paid to the tenets of liberalism.[30] Protectionism remained the knee-jerk reaction of those in the public and private sectors alike.

The importance of such a claim in the context of the present study is particularly great in that the environment of the 1970s and 1980s was one which threatened the economies of Britain and France in a very stark way. Both countries found themselves locked into a series of international commitments to a liberal economic order via the EEC and more widely, the General Agreement on Tariffs and Trade (GATT), which limited their scope for manoeuvre. At the same time, both had economies which were sufficiently open to feel the effects of interdependence with other economies. They were therefore obliged to reassess their commitments and the impact of interdependence. Given the character of the French economic tradition, it should not be at all surprising to hear that "the French state attempted (through protective measures) to control and direct the terms of interdependence",[31] whereas Britain remained more firmly linked to international commitments on open and free trade, despite the growth of protests from those domestic industries suffering from the recession.

Britain and France are seen therefore to differ sharply in their styles of policy making and in their economic orientations. In both areas, the role of the state is very much more important in the French than in the British context. It is for this reason that France can be characterized as a state-led society, whereas Britain resembles more closely a society-led state. To underline the contrast it is useful to refer to the two competing attitudes towards the relationship between state and society that Berki identifies. These he entitles _transcendentalism_ and _instrumentalism_ and he defines them as follows:

> "Transcendentalism ... refers to the belief that man primarily belongs to a _moral_ community ... that the community has a paramount moralising function and is, therefore, logically speaking 'prior' to its members. Individuals, it is held, are united together in the service of common and moral goals ... instrumentalism, by contrast, embodies the belief that man primarily belongs to an _interest_ community, in other words that the group's existence and functions are external to individuals and are not directly related to their moral feelings and aims. Individuals become or continue as members of associations because they see that the latter as 'instruments' promote their own individual private aims"[32]

In France the idea of the state is critical as the expression of collective rather than particular interests and in downgrading those particular interests through an activist policy-making elite concerned to guarantee a high level of protection for the domestic economy. For this reason the state can be seen to represent a set of broad moral principles, justifying a 'transcendentalist' interpretation of the French political process. By contrast, an 'instrumental' conception of British attitudes can be accepted because of the weakness of the idea of collective state principles in guiding individual action. Both the style of policy-making and the dominant economic outlook underline the importance accorded to particular interests, with the state's role limited to one of arbitration rather than direction. We will use this general contrast between the two countries to understand theirresponses to change in the fisheries sector.

3. Evaluating the response to change

This study seeks to go beyond an explanation of the response of the two states to change in the fisheries sector: it also attempts to evaluate how successful they have been. Such an evaluation seems to the author to be important for two reasons. Firstly, the plight of the industry, especially in Britain, has generated an enormous volume of invective, combined with a high level of frustration and helplessness. A not untypical reaction was that of Sir Leo Schultz, the Mayor of Hull, when he testified to the Trade and Industry Subcommittee of the House of Commons.

> "... surely there is somebody who can see through this tangle of causes and effect, which has caused this dilemma and who can find for us the answer to this problem."[33]

Even if there is no single answer, any study of the area has a responsibility to address the issue of succes.

The second reason justifying an assessment of success is that within Britain, there have been a number of writers who have argued that the British industry would have benefitted or indeed would still benefit from adopting elements of French organisation and policy.[34] It will be suggested here that these comparisons fail to recognize the complexity of the differences between the two states, and that they therefore demand a competing assessment of the two countries' response to change.

This study will not apportion praise or blame in relation to particular policies. It will not try to suggest, for example, that Britain's fishing industry would have avoided difficulty if the country had not belonged to the EEC. Rather it will be argued that Britain and France had clearly contrasting priorities which were reflected in the way interests were balanced. However, neither way seems to the author to be free of difficulty: both pose hard choices, especially where the straightforward transfer of practices is not possible. The choices to be made can be identified in two particular areas.

The first issue involved in the balancing of interests is the identification of the relative importance of the various sectors <u>within</u> the fishing industry. This study concentrates its attention on the catching sector but it cannot be assumed that governments necessarily did do or should do so. There are a whole series of activities, both upstream and downstream, whose interests may not coincide with those of the catching industry. The most obvious example is the relationship between catchers and merchants. The latter are not necessarily dependent, within one country, upon the produce of the former. As Mr. Ellington, the

President of the Hull Fish Merchants' Protection Association explained in his testimony to the Trade and Industry Subcommittee:

> "We are rather like the oil people whose oil wells run out. We have no control over our own raw material. It is up to us to try to get our supplies where and when we can. We have no objection to importing fish from any place." [35]

The possibility of such imports necessarily creates a tension between merchants and indigenous fishermen. Equally, such a possibility underlines the need to place any idea of success in a wider context than that defined by the catching sector of the industry. Should government policy be geared to maintaining an industry as close as possible to its traditional shape? Or should it be prepared to subordinate this objective to the demands of the consumer for the product he or she wants, at the best price? It will be argued here that the behaviour of the British and French governments illustrates that they gave rather different answers to these questions and were faced with different kinds of difficulty as a result.

The second area of choice concerns the treatment of the interests within the catching sector itself. There are not only divergences between fishermen and other sectors of the industry, but also between different parts of the catching sector. The difficulty of overcoming these individual interests inside the catching sector in pursuit of a common interest is an important further theme of this study. In particular, it will be argued that fishermen in both countries faced, in a variety of forms, the 'free-rider' problem of providing a 'collective good' when the contribution of a single fisherman to that good makes little or no difference and he therefore has no incentive to make that contribution. Under such circumstances, as Olson has argued:

> "...it is certain that a collective good will not be provided unless there is coercion or some outside inducements that will lead the members of the large group to act in their common interest".[36]

As in the case of the relative importance of the catching sector, the two governments concerned took distinct positions. In this case they perceived differently the degree of 'coercion' and 'outside inducements' that could be used so as to encourage the pursuit of the common interests of the sector.

These differences will themselves be set in the wider context of a balance between the articulation and the aggregation of interests. It will be suggested that there was an inevitable tension which any assessment of success cannot ignore.[37] The more that particular interests were allowed to make their own complaints known, the more difficult it was to identify common interests. To prolong discussion in the interests of the widest ventilation of grievance was effectively to veto decision. By contrast, to push through a decision in order to move beyond debate to action risked undermining consensus. The more intense the pressure for collective solutions, the less the opportunity for differences to emerge and for the aggrieved to feel that they had had their say. Again the broad attitudes towards the state in the two countries will be seen to have influenced the way in which this balance was struck.

Finally, in this section, it is important to underline that any notion of success has to be set in the context of the changing character of the issue. In the period under discussion, it was no longer possible to treat the foreign and domestic aspects of the fishing industry as belonging to separate compartments. The boundary between the two aspects became increasingly blurred especially as the debate moved into the context of the EEC. Thus the commitments of governments to the maintenance of employment and to the promotion of regional development at home were not easily disposable once discussion was under way in Brussels. By the same token, the commitments involved in membership of an organization like the EEC, such as the free movement of goods, reduced the scope for a policy of import controls, which a threatened domestic interest, like the fishing industry, might call for. The resultant interlinkage between the countries of the EEC can be said to make it increasingly irrelevant to talk of national success: what counts is their joint ability to reach agreement.

However, this study will argue that whatever the level of integration and whatever the need for further integration, national policies and attitudes persist and are important in determining the nature of the integration achieved. Thus contrasting national stances on the relative importance of the catching sector or of different groups within that sector can be detected in the debates which take place at the EEC level. In this sense the 'acquis communautaire', the shape of the Community's policies, is in constant flux and renegotiation.

At the same time, the difficulties that the separate states and industries have in generating collective interests out of individual ones are reflected in Community discussions. Despite the agreement of January 1983, success for the integration process of the EEC in the fishing area remains elusive for the same 'free-rider' reason that individual states face domestically. Why should national industries accept to contribute to the collective good, when their individual contributions are not critical and when they suspect that their competitors will not follow suit? And yet unless this dilemma can be solved, both nation states and the Community may face the unpleasant prospect of decreasing catches and declining catching sectors. In this sense national success cannot ultimately be separated from success in a wider forum.

4. The method of investigation

The explanation and evaluation of change in the fishing industry will be developed here in two ways: firstly, the relationship between the government and the industry will be examined using a set of four perspectives; and secondly, each perspective will be considered on the basis of how we might expect the relationship to have developed in the light of the material in the first and second sections of this chapter. On the one hand, the changes in the issue area in terms of resource scarcity, resource ownership and the economic conditions of exploitation will be taken and their anticipated effect on the relationship considered; on the other hand, the contrasting character of relations between state and society in the two countries will be used to prompt discussion of the likely development of the relationship.

The use of perspectives is intended to underline that the relationship between interest groups and governments needs to be conceptualized in different ways and should be looked at in terms of the range of behaviour that they display towards one another. The categories used here were derived by the author from the work of Finer. He suggests that four main views have been presented as to the relationship which ought to exist between groups and government: groups as opponents of the government, groups as substitutes for government, groups as extensions of the government and groups as intermediaries between public and government.[38] These views are not just useful as normative categories: they also offer a way of dividing up the observed behaviour of groups and government. Here they have been renamed and reordered to generate a continuum, ranging from the closest form of involvement to the remotest connection between the two.

The first perspective is entitled 'interventionist' and concentrates its attention on the direct involvement of the government in determining the behaviour of the industry. Unlike the other three perspectives, it is a 'top-down' view but one where the industry responds to as well as reflecting the government's involvement.

The second, 'mediatory' perspective sets the relations between industry and government in a broader context. It considers the way in which the industry's representatives transmitted the views of its members by lobbying government. Here the opportunity arises for going beyond direct contacts with government and appealing to other sources of influence in order to obtain a better deal.

The third, 'direct action' perspective looks further than traditional mechanisms of influence and incorporates behaviour which goes outside the legal framework linking the two sides. The assumption here is that the 'traditional mechanisms' are no longer effective and that the industry feels it has no other way of making its voice heard.

The fourth, 'self help' perspective directs attention away from the search for influence and towards the efforts of the industry to organize its economic future by itself. In principle, such self-help enables an industry to avoid dependence on government and thus to ensure maximum distance from it.

The purpose of these four perspectives should be clarified at the outset. Above all, it is important to stress that they are aids to organization rather than models of behaviour to be shown to offer better or worse 'fits' in terms of explanation. From this it follows that the intention is not to establish that one of the perspectives represents the 'normal' relationship between industry and government, the others applying to divergences from the norm.[39] It is true, but for the purposes of this study uninteresting, to point out that in terms of frequency fishermen were not resorting to direct action as regularly as their leaders were transmitting their views to government. Any search for 'normalcy' seems to the author to risk relegating certain kinds of behaviour to the realm of the aberrant, and to minimise the importance of occasional events.

More important, to attempt to create a hierarchy among the perspectives detracts from one of the tasks of this study, namely, to underline the relationships among them. In particular, the study is concerned to reveal that behaviour discussed under one perspective may help to explain behaviour discussed under another.Thus it will be argued here that much can be learned from the character of intervention about the nature of direct action in the two countries. Similarly, the type of self-help developed will be presented as strongly influenced by the shape of both the intervention and the mediation that took place. In this sense, it will be claimed that though behaviour relevant to all four perspectives can be identified in both countries, the differences between them suggest that the behaviour followed a distinct logic in the contrasting national environments. It is these logics of action that will be explored rather than the possibility of identifying any norm in the relationship between the fishermen and their governments.

In addition to developing the four perspectives, each of the chapters concerned is linked to two related claims as to how we might expect the relationship between the fishing industry and government to develop given the severity of change within the issue area and the general character of relations between state and society in the two countries. In all four cases, it is assumed that change in the issue area would be likely to push the relationship in the same direction in both countries but that the nature of the differences between the two countries would accentuate the development in one of them.

Within the interventionist perspective, we will start from the assumption that the impact of change would be that the industry would demand increased intervention from government to protect it and that the government would feel itself obliged to cater for those demands. At the same time, we will suppose at the outset that the 'dirigiste' French tradition and the British aversion to intervention would mean that government efforts to determine the shape of the industry were more developed in France than in Britain.

The mediatory perspective will take as its point of departure the idea that the representatives of the industry would be obliged to transmit the views of the industry through a much wider set of channels because of the disgruntlement of their members at the effects of the changing environment. It will also suggest that the higher French acceptance of the legitimacy of public action and the greater elasticity of British political life would make such developments more widespread in Britain than in France.

In discussing the direct-action perspective, we will argue that the expected result of major change in the issue area would be for the industry to be pushed into forms of direct action in as far as it perceived that its demands for assistance made no headway through the conventional channels. Further we will anticipate that the importance of the search for consensus in British policy-making and the 'limited authoritarianism' implicit in the French activist policy style would make such direct action more common in France than it was in Britain.

As far as the self-help perspective is concerned, the premise will be that the industry would respond to change by intensifying its efforts to help itself. In addition the expectation will be that the difference between the character of French and British state intervention in the economy would make such self-help efforts much less individualistic in France than they were in Britain.

5. Conclusion

This chapter has presented a general outline of the study. It is, first of all, a study of the response to change within the fishing industry in Britain and France. That change cannot be dated precisely, but consideration of its effects will be concentrated on the period beween 1975 and 1983. The widespread protests of the fishermen in both countries in the earlier year and the conclusion of the CFP in the later provide convenient episodes for delimiting the main period for investigation. Equally, as indicated earlier, it is the changes in the catching sector of the industry which are at the centre of attention: the rest of the industry will be considered only in as far as its fortunes impinged directly on those of the catchers. In seeking to provide an understanding of the extent of these changes, the next chapter will examine the way in which the structure of the catching sector of the industry changed in the two countries under the pressures of resource scarcity, resource reallocation and the change in economic conditions.

The study is also concerned to show how the content of the fisheries' issue changed for the two governments. From being a limited, sectoral concern, it was transported into a broader, more highly-politicised environment where Britain and France were both subject to the constraints and opportunities of a collective search for a solution to the problems that the industry faced. This changed environment, based on the two countries' membership of the EEC, and their differing responses to that environment will be the subject of chapter three.

Chapter four draws together the government and industry and brings out the next theme of the study: that the response to change can only be fully understood in the context of the general relationship that has grown up between the two parties in both countries. The chapter will introduce the topic by looking at the very different institutional arrangements in Britain and France and the contrasting ways in which they were challenged and adapted in the period under consideration.

The following four chapters will develop the contrast between the two countries in the context of the four perspectives, introduced in the previous section. Thus chapter five will examine the interventionist perspective, chapter six the mediatory perspective, chapter seven the direct-action perspective and chapter eight the self-help perspective.

Finally, chapter nine will return to the major themes that run through the whole work. The distinction between France as a state-led society and Britain as a society-led state will be reviewed in the context of the evidence generated by the study. On the one hand, the chapter will consider the way in which the relationship between industry and government developed; on the other, it will underline the dilemmas involved in judging the two states as more or less successful in responding to change in the fishing industry. The British and French responses were different, but neither provides an unproblematical answer to the difficulties of a contemporary fishing industry in Western Europe.

2 The impact of economic change

1. Introduction

1975 was widely regarded as disastrous for the fishing industries of Western Europe. The Organisation for Economic Cooperation and Development (OECD) used the term, "catastrophe" to describe a situation where the catch levels of member states were down by 500,000t. on 1974. In France, Dubreuil, the President of the Central Sea Fishery Committee (CCPM) spoke of "the most complex and most tragic" crisis in the memory of any fisherman[2] and in Britain, Wood, the head of the Aberdeen Fishing Vessels Owners' Association, commented at the beginning of 1975 that "the last year had seen the most incredible turn-round in the fortunes of our industry."[3]

Such sombre talk was to predominate within both the French and British fishing industries throughout the rest of the 1970s and only in the first years of the 1980s did the tone begin to change. In France 1981 was described as a "year of recovery"[4] and later in Britain, despite some misgivings, the paper of the industry conceded that the conclusion of the CFP in January 1983 put "an end to many of the uncertainties which have wrecked British fishing since we joined the EEC."[5]

This chapter will explain why 1975 heralded such a difficult era: first, it will stress certain basic features of the industry which had exposed fishermen to recurrent misfortune before 1975;

second, it will suggest that what was new in the 1970s was the internationalization of the sector, a process which significantly increased the vulnerability of fishermen to economic change; and third, it will point to the different effects that this new vulnerability had on the British and French industries.

2 The character of the fishing industry

It is not possible to understand the significance of the events of the 1970s without an awareness of the peculiar characteristics of the fishing industry as an economic activity. In certain respects, as a primary industry, it resembles agriculture. The weather can play an important part in disrupting activity and the cost of equipment, outside the producer's control, is a major drain on his resources. However, the fisherman is subject to an even greater degree of uncertainty than the farmer. First of all, when he sets sail, he can never be sure how much he will catch. Secondly and more importantly, he can have only a limited idea of the earnings that his catch - large or small - will bring on his return to port. This is because fishing has been predominantly governed by the law of the auction, where merchants compete with one another to buy fish at the price which they consider the consumer will pay. If the sale price in the shop is too high, the consumer will turn to competing foodstuffs, notably meat. On the other hand, the lower the sale price, the less the fisherman will have earned. The reason for this is that the auction price and the sale price of fish are generally acknowledged to remain in roughly the same ratio. Thus in France, the latter is said to be three times the level of the former, while in Britain the difference has been calculated at 400%.[6] The nature of distribution and commercialisation make it very difficult to modify these relationships.

The fisherman is therefore dependent upon the behaviour of those further down the economic chain from him. If he returns with too high a catch or more likely, if the catches of those ships returning on a particular day are too high, then prices will go down and there is little he can do about it. Given the extreme perishability of fish he is not free to wait until another auction before putting the product on sale, at least if he is selling fresh rather than frozen fish. At the same time, the lower the catch, the less likely he is to make a profit after covering his costs, in terms of fuel, payments to the crew, repayments to the bank, etc. These costs may go up but the rise has no influence on the price that the merchant or the consumer pays. This is not to say that fishermen can never benefit from the auction system: they can and often did even in the period under discussion. Nevertheless, they are in a unique situation which was summed up by Parres, the Secretary General of the Union of French Fishing Boat Owners (UAP), as follows:

> "Fishing is the only economic activity in which there is no relationship between the cost price and the sale price of the product. The reason is that the fish merchants ply their trade in a food market which is heavily controlled and where substitute foods and imports prevent the passing on of costs in an economic way".[7]

In addition to these problems of profitability, fishermen had also been aware for a long time that overproduction could threaten the stocks of fish from which they derive their livelihood. Already in 1946 the countries of the North East Atlantic had signed an Overfishing Convention, conscious of the effects of indiscriminate catches in the pre-war period and eager to benefit from the stock recovery that world war had engendered. La Rochelle was an example of a port where this recovery was obvious: its catches of hake in 1946 were more than double what they had been in 1938.[8]

However, the level of catches continued to rise throughout the post-war period and this led to periodic slumps, such as those which hit Britain and France in the late 1950s and early 1960s. As prices dropped, ships were kept in port and governments provided subsidies to keep the fishermen in business until the situation improved. And improve it did, for the 1960s witnessed dramatic increases in efficiency. In Britain, for example, average, annual landings of those ships that sailed to the most distant grounds rose from 29,900 cwts. in 1963 to 42,300 cwts. in 1969.[9]

The factors which enabled these improvements to take place are much clearer now than they were at the time. First of all, the principle of the freedom of the seas for fisheries remained the dominant principle in relations between the states of the North Atlantic. It is true that Iceland laid claim to twelve mile fishery jurisdiction during the first 'Cod War' with Britain between 1958 and 1961 but the situation appeared to be settled by the so-called London Convention of 1964. Under this agreement European states acquired the right to reserve fish stocks up to six miles from their coasts exclusively for their own vessels, with the nationals of other signatories guaranteed access between six and twelve miles to the extent that they had traditionally fished there.[10] As a result it continued to be possible for new grounds to be developed as well as for the old grounds to be maintained, without apparently seriously threatening stock levels.

Secondly, it proved possible to increase productivity without a corresponding increase in costs. New types of ship, notably the stern trawler which came to Britain in 1961 and to France in 1965, combined with new techniques, in particular the use of sonar for locating shoals, to make fishing more cost-effective and less labour-intensive. Indeed the fishing industry appeared to be following the path of agriculture with marginal producers leaving the sector and the large-scale concerns assuming an ever greater role. Thus between 1960 and 1972, the number of fishermen fell in Britain from 28,254 to 22,703 and in France from 50,670 to 34,609. Over the same period, production rose in France from 626,764 t. to 670,143 t., while in Britain the catch went up from 813,382t. to 954,837 t.[11]

The fact that the good times of the 1960s were premised on the maintenance of an open-seas regime and of a situation where costs did not outstrip productivity was much less evident at the time than it is now. There was a general feeling that the industry was governed by economic and biological cycles: prices might fall, stocks might decrease but eventually, the fish would return, eventually, the market would recover.[12] Indeed the chance character of the auction symbolized this inherent confidence that the next landing would be the one that brought a bumper wad of notes to the fishermen. "The auction", as one French writer has put it, "is the temple of fish".[13]

By a remarkable irony the critical developments of 1975 and after followed a period when the "temple of fish" was giving the fleets in both countries rewards that were greater than anything previously known in the industry. In Britain the general level of profitability was so good that in July 1973 the government withdrew operating subsidies without causing any major disquiet. Indeed some saw this as a sensible move, as fishermen made a "mad rush for new boats", reinvesting their profits to ward off high tax bills and overseas competition.[14] In France, too, between 1971 and 1975 there was a major renewal of the fishing fleet, aided by state subsidies. Nearly one half of the largest ships were replaced, with 130 new boats coming into service, and amongst the smaller vessels, over 250 were built with the help of grants and loans. In Southern Brittany alone 60 new trawlers of over 100 feet in length were delivered during this period.[15]

Nowhere perhaps did the future look better than on Humberside. Both Grimsby and Hull were making annual catches of over 200,000t. which together accounted for between one third and one half of the British total. No ports in Europe came anywhere near to matching such catch levels: in France, for example, only at Boulogne were annual landings above 100,000t. The start of the second 'Cod War' with Iceland in September 1972 and attempts by the North East Atlantic Fisheries Commission (NEAFC) to introduce quotas to restrict catches appeared no more than minor problems.[16] Yet by 1982 the combined catches of Hull and Grimsby at less than 60,000t. amounted to only a little over half of the catch at Boulogne.[17] Why did this happen and why did the fishing industry as a whole undergo structural rather than cyclical change?

3 Fisheries and the international economy

The basic difference between the earlier difficulties that the fishing industry had faced and those it faced in the 1970s was that the dynamic of cyclicality became less important than the dynamic of externality.[18] Whereas previously the fluctuating fortunes of national fishing industries were chiefly determined by the behaviour of those within the industry, it now became clear that their future shape depended on events within the international economy as a whole. The contours of the international environment had remained stable for a long period but during the 1970s the link between the national and international environments emerged as a critical determinant of the prosperity of the fishing industry.

The first overt illustration of this new reality was the dramatic increase in the price of fuel for fishing, prompted by the actions of the OPEC oil producing states at the end of 1973 and the beginning of 1974. As a result one gallon of fuel which had cost 7 pence in Britain in September 1973 cost 21 pence in March 1975, while in France, a price of 12 centimes a litre in June 1973 rose to 42 centimes in January 1974.[19] The possibilities of assimilating such a rise by reduced petrol consumption were relatively limited. Even before the autumn of 1973 fuel was calculated to represent 10% of the running costs of the French fishing fleet and that figure now rose to 25%.[20]

Nor was it only a question of fuel: the effects of the rise were felt in all the material necessary for the industry from nylon nets to the vessels themselves. Purse-seine net equipment which had cost £13,000 in 1967 went up in price to £80,000 in 1974 because of the fuel base needed to make it.[21] And as shipyards were obliged to increase prices or go bankrupt, a British fisherman could be obliged to find nearly £450,000 in 1976 for a boat ordered in 1972 at just over £130,000.[22]

These difficulties were compounded when the OPEC states increased the price of oil again in 1979. By 1980 French fishermen were paying 1.23FF per litre, ten times more than the price in early 1973.[23] At the same time, the average price of fish in France did not increase at anything like an equivalent rate: whereas in 1972, 1000 litres of fuel could be bought by the sale of 45kg of fish, by 1982 between 180 and 200kg were needed to make the same purchase.[24] Again this was not an isolated national phenomenon but one faced by all the Community countries, including Britain, and there too the fuel price rose to similar levels to those in France.[25] As we shall see, the question of exactly how similar those levels were became an important political issue.

The second indication of the changing relationship between the national and international environments for the fisheries industry emerged at UNCLOS III. The European states traditionally supported the idea that fishing should be open to allcomers except within a relatively narrow coastal band, a concept enshrined in the 1964 fisheries convention. But by 1975 it was becoming obvious that the international community as a whole was turning towards the idea of a 200 mile EEZ, within which the coastal state would have exclusive rights to the exploitation of resources, including fish. What subsequently happened was that the traditional European attachment to extensive freedom to fish was traded off within the Law of the Sea conference for the rights of the warships and merchant ships of the industrialised world to pass through the straits of third world states.[26] In view of these wider developments, the decision of Iceland to declare a 200 mile fishery zone from October 15 1975 should be seen not as an isolated act but one that anticipated the imminent action of coastal states throughout the world.[27]

The effect of this change upon the fishing industries of Western Europe was very dramatic. It eliminated the automatic right to fish in the prolific waters of Canada, Iceland, Norway and the Soviet Union. On November 30 1976, for example, after a final 'Cod War' with Iceland, British trawlers were excluded for ever from grounds which they had been fishing since the 1880s and where they had been catching 130,000t. per year since the settlement of the previous conflict in 1973. All in all, it was calculated that the losses in third-country waters amounted to

230,000t. for Britain, 173,000t. for West Germany and 52,000t. for France. 28

The gravity of the situation was not as great as it might have been, because from January 1977 the members of the EEC agreed to extend their fishing limits in the Atlantic simultaneously to 200 miles. As a result, third-countries, notably the Eastern Europeans, were themselves excluded from this zone, unless they were willing and able to offer some form of reciprocal access. However, the benefit was only a limited one for two reasons. Firstly, those ships which were suitable for fishing at long distances from European ports were not the kind of ships needed for fishing closer to home. They consumed too much fuel and they could not make the kinds of catches they had made in the distant northern waters. When they attempted to do so they not only came into conflict with other smaller vessels but also placed more strain on the stocks within the Community zone. The result was a severe level of overcapacity, a case of too many boats chasing too few fish. Many of the larger ships were therefore laid up.

The second reason why the creation of an EEC zone offered only limited benefit was that the spread of the EEZ concept gave those countries, like Canada, from whose waters EEC states were increasingly excluded, the opportunity to develop their own fleets and to export surpluses to the Community. Moreover, the fact that their catch rates were higher, and their fuel costs per unit of fish lower, meant that they could export at prices which undercut those of Community fishermen.

The possibility of taking advantage of the liberal world economy in this way was something already acknowledged before the 1970s. The spread of freezer trawlers in the 1960s offered a means of overcoming the problem of the perishability of the product: frozen blocks of filletted fish could be transported between countries and continents, thereby posing a severe challenge to the domestic fresh fish market. This was a particularly interesting proposition for those large, multinational companies, who could use economies of scale to take advantage of the freezer revolution with or without a catching capacity of their own. Unilever, for example, the world's largest food multinational, has a major shareholding both in Bird's Eye in Britain and Nordsee in Germany.[29] However, while the German firm has its own ships and shops, the British company has never had any vessels of its own, always relying on deals with trawler owners. Either way Unilever could assure itself of supply for its downstream food activities.[30]

But the increasing integration of fishing into the world economy was not something that the frozen fish market and multinationals alone made possible. There was the further question of the relationship between fish prices and those of other foodstuffs on the international market. One of the effects of the recession in America after the increase in oil prices in 1973 was that prices of meat and poultry fell. This in turn led to a drop in consumption of fish which drove the traditional exporters to America to look for new markets. These they found in the Community, where the market was overwhelmed by such unlikely candidates as Argentine hake. As a result prices plummeted and few mechanisms were available to do anything about it.[31]

It is in this context that the spread of 200 mile fishing zones has to be seen. They offered a further incentive for fish-exporting countries to turn their attention to the Community. By 1978, Canada was able to become the world's premier exporter of fish with a fifth of its trade going to the EEC. At the same time, the Community countries, with the exception of Denmark, Holland and Ireland, all became more dependent on imports. Between 1970 and 1982, British imports more than doubled in quantity though total supplies (i.e. landings plus imports) remained roughly the same, while in France between 1975 and 1979, the value of imports rose from 75% to 115% value of landings by French vessels. French vessels were no longer able to make catches that were worth as much as the value of imports.[32]

The effect on individual fishermen was dramatic and showed how the future of the industry could not be assured simply by hard work and dedication. In August 1975, for example, the brand new freezer trawler, 'Arctic Galliard', returned to Hull with a catch of 845 tonnes, the largest that any British ship had ever made. However, it made a loss of £33,000, when fuel and depreciation costs were set against the price that obtained at the auction in a market heavily influenced by imports.[33] Nor was this an isolated example. Fishermen were faced with the possibility of their catch being converted for fish meal, a possibility which when added to the dangers of catching fish off Iceland during the final 'Cod

War', make it hardly surprising that some asked themselves whether it was all worth it:

> "We leather ourselves risking life and limb off Iceland to catch this lot, with gunboats sat on our tails, then come home to see hours of graft and worry ending up as meal."[34]

Admittedly the dangers of Icelandic gunboats were not ones that many fishermen had to face but all were obliged to be concerned about the price that their catch would fetch, not just because of the workings of national factors but because their industry had undergone structural change and was being increasingly integrated into the international economy.

4. The British and French fishing industries

Up to this point it has been assumed that the subjects of this study, namely the British and French fishing industries, could be treated as similarly structured and similarly affected by the changes of the 1970s. It is certainly true that these two industries resemble one another more than either resembles any other fishing industry in the EEC. By the beginning of the 1980s, both had about 20,000 fishermen, around 0.1% of the total workforce in each country;[35] both were landing over 700,000 tonnes per year, catches representing about 15% of the Community total in each case;[36] and both had nearly 500 vessels of over 100 tonnes balanced by a substantially greater number of smaller vessels.[37] However, these similarities conceal substantial differences both in the shape of the industries and in the impact of change upon them. To illustrate these differences this section will look in turn at where the two countries fished, who undertook the fishing, what species they caught and from where they set out.

4.1 Catch areas

Although the two countries were catching similar amounts of fish by the 1980s, ten years earlier the situation had been quite different. At that time the British catch amounted to nearly 1,100,000t. whereas France was still landing around 700,000t.[38] It has already been suggested (p 5) that the introduction of 200 mile zones affected Britain more seriously than it did France and the difference in third country losses helps to explain why their catches developed in such contrasting ways. To understand what happened more fully it is useful to look at Figure 1, which shows the areas covered by NEAFC and to compare it with Table 1.[39] The table shows how much Britain and France caught outside the Commmunity 200 mile zone before the final 'Cod War' and after the extension of limits at the beginning of 1977 (excluding a very small French catch in areas 1X (Portugal), X (Azores) and X11 (North Azores).

FIGURE 1 - North East Atlantic fishing areas

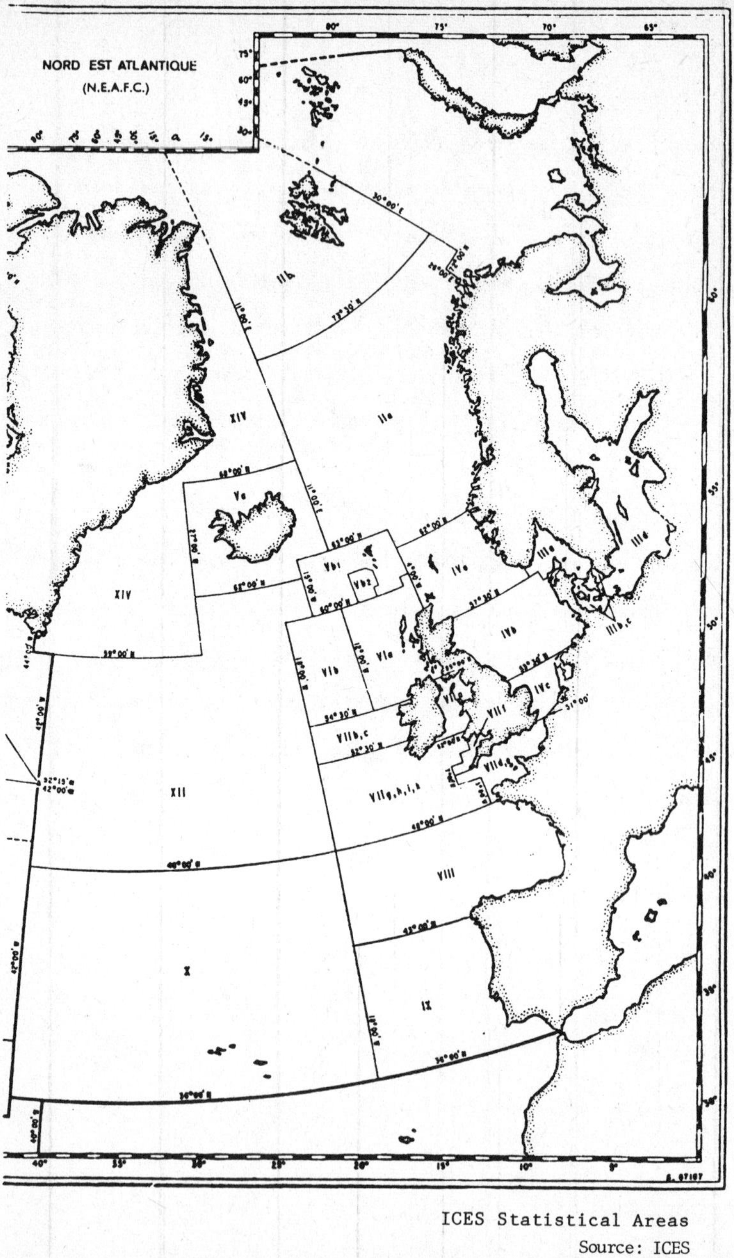

ICES Statistical Areas
Source: ICES

TABLE 1

British and French landings from selected NEAFC areas in 1974 and 1977 (in tonnes)

	1974		1977	
	BRITAIN	FRANCE	BRITAIN	FRANCE
Barents Sea (I)	96,313	11,985	58,013	8,341
Norwegian Sea (IIa)	30,080	15,221	32,237	13,304
Spitzbergen and Bear Island (IIb)	14,366	30,686	30,691	2,182
Iceland grounds (Va)	142,034	248	29	-
Faroe Plateau (Vb1)	31,759	25,194	21,291	29,545
Faroe Bank (Vb2)	4,432	-	219	-
TOTAL	318,984	88,334	142,480	53,372

The figures illustrate very clearly the impact of the Icelandic success in establishing a 200 mile limit: in 1977 the total British catch in non-EEC waters only slightly exceeded her catch in 1974 in Icelandic waters alone. The French, by contrast, sustained a much smaller loss: in 1974 the distant grounds had together contributed only about 13% of the total catch in NEAFC areas (over 650,000 t.) and the percentage fell relatively little to around 9% in 1977 when the total was just under 600,000t. In the British case, although the total NEAFC area catch remained at close to a million tonnes, the percentage from the distant grounds declined in the three years from over 30% to about 14%.

The difference in impact of the Icelandic move needs also to be understood in terms of both the balance between NEAFC area catches and non-NEAFC catches and the catch distribution within the NEAFC areas that form part of the EEC 200 mile zone. Although France did see a decline in her catch in the distant water areas of NEAFC, she continued to make substantial landings from areas outside the North East Atlantic. In 1976, for example, nearly 20% of all French landings came from catches made in the Mediterranean (over 50,000t.), the West African coast (over 60,000t.) and the North West Atlantic (over 35,000t.). Moreover, despite the spread of 200 mile EEZs, the overall contribution of these three areas was at a very similar level in 1982.[40] Britain, by contrast, has never been active outside the NEAFC area (beyond a small catch off Canada) and the move to extended fisheries zones did not induce her to change this practice, as we shall see in Chapter Five.

For both countries, however, the largest share of their fishing took place and continues to take place within 200 miles of the coasts of the EEC. In 1976, for example, over 40% of the British catch in NEAFC areas c areas came from the North Sea alone and over 25% of the equivalent French catch from the Bay of Biscay .[41] What differentiates the two countries is that whereas British fishermen rarely move outside the UK part of the Community zone, their French counterparts have always depended to an important extent on access to the waters of other EEC states, in particular Britain and Ireland. The importance of this difference cannot be overemphasized in that it lay at the heart of much of the dispute that raged within the EEC over the shape of the CFP.

When the prospect of a joint extension of fishing limits to 200 miles in the EEC loomed, the French Economic and Social Council (CES) commented:

> "The zones within the Community will remain, as a matter of principle, open to the fishing fleets of all the member states: that they should do so is absolutely essential for France, which takes 72% of her catch from Community waters alone."[42]

This presence, however 'essential', looked very different from the British point of view: the country appeared to be surrounded by the vessels of foreign states. The Trade and Industry Subcommittee of the Expenditure Committee was presented in 1976 with maps which underlined the activity of non-British ships around the coasts of Britain and the lack of activity of British ships in the waters of other EEC states.[43] Figure 2 shows that the French, for example, were fishing in 1974 all around Britain in the English and Bristol channels, the Irish Sea, the waters off North-West Scotland, near the Shetlands and in the southern North Sea. At the same time, Figures 3(a) and 3(b) point to the chief effort by British vessels for pelagic and demersal species (i.e fish swimming in the water column and those near the seabed) as taking place on the British side of a median line equidistant between Britain and the surrounding states. Even on a very liberal estimate, the level of the British catch off France can hardly have amounted to more than 1000t. for the year in question. The paucity of the grounds off France offered little incentive for British fishermen to visit them, either then or later.

Chart 216

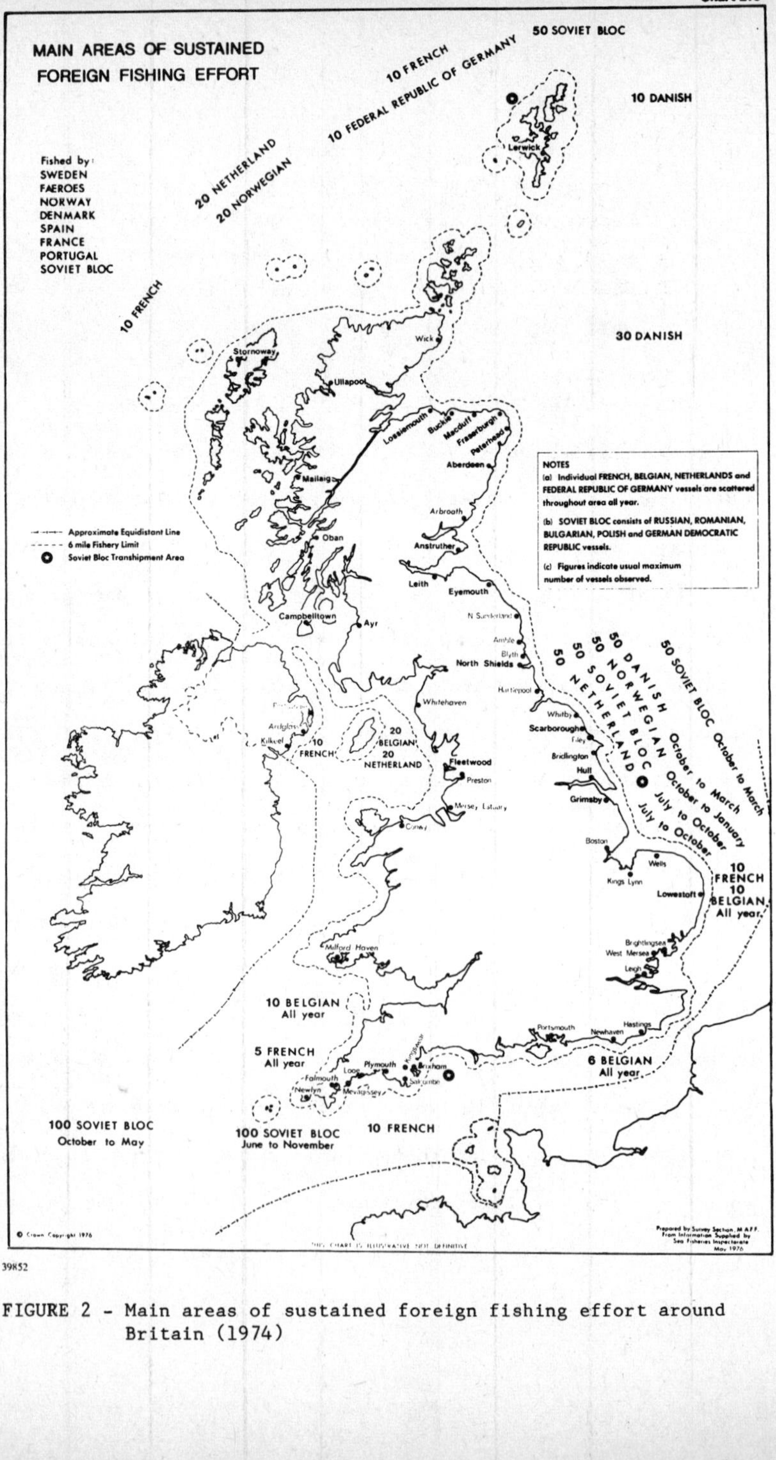
MAIN AREAS OF SUSTAINED FOREIGN FISHING EFFORT
Fished by:
SWEDEN
FAEROES
NORWAY
DENMARK
SPAIN
FRANCE
PORTUGAL
SOVIET BLOC
50 SOVIET BLOC
10 FRENCH
10 FEDERAL REPUBLIC OF GERMANY
10 DANISH
20 NETHERLAND
20 NORWEGIAN
10 FRENCH
30 DANISH
Approximate Equidistant Line
6 mile Fishery Limit
Soviet Bloc Transhipment Area
NOTES
(a) Individual FRENCH, BELGIAN, NETHERLANDS and FEDERAL REPUBLIC OF GERMANY vessels are scattered throughout area all year.
(b) SOVIET BLOC consists of RUSSIAN, ROMANIAN, BULGARIAN, POLISH and GERMAN DEMOCRATIC REPUBLIC vessels.
(c) Figures indicate usual maximum number of vessels observed.
50 SOVIET BLOC October to March
50 DANISH October to March
50 NORWEGIAN October to January
50 SOVIET BLOC July to October
50 NETHERLAND July to October
10 FRENCH
20 BELGIAN
20 NETHERLAND
10 FRENCH
10 BELGIAN
All year
10 BELGIAN
All year
5 FRENCH
All year
6 BELGIAN
All year
100 SOVIET BLOC
October to May
100 SOVIET BLOC
June to November
10 FRENCH
Lerwick
Stornoway
Wick
Ullapool
Lossiemouth
Buckie
Macduff
Fraserburgh
Peterhead
Aberdeen
Mallaig
Arbroath
Oban
Anstruther
Leith
Eyemouth
Campbeltown
Ayr
North Shields
Whitehaven
Hartlepool
Whitby
Scarborough
Filey
Bridlington
Ardglass
Kilkeel
Fleetwood
Preston
Hull
Grimsby
Conwy
Boston
Wells
Kings Lynn
Lowestoft
Brightlingsea
West Mersea
Leigh
Milford Haven
Portsmouth
Newhaven
Hastings
Plymouth
Looe
Falmouth
Newlyn
Mevagissey
Brixham
Salcombe
© Crown Copyright 1976
Prepared by Survey Section, M.A.F.F. From Information Supplied by Sea Fisheries Inspectorate May 1976

39852

FIGURE 2 - Main areas of sustained foreign fishing effort around Britain (1974)

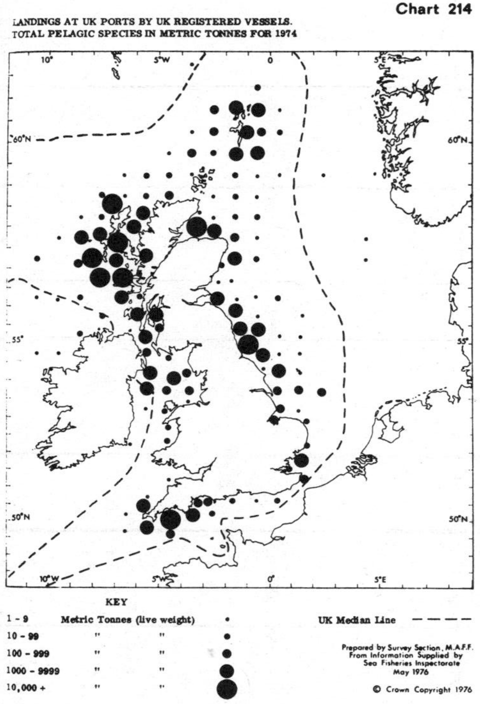

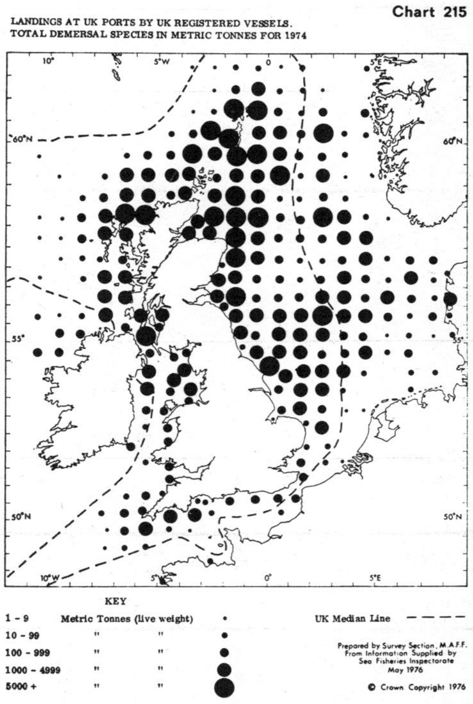

FIGURES 3a and 3b - Location of demersal and pelagic catch by British vessels (1974)

After the move to 200 mile limits, less was made in Britain of the presence of foreign vessels (the East European were in any case shortly to leave) and more of the contribution in terms of volume of fish that came from within the British part of the so-called EEC 'pond'. As Table 2 indicates, nearly 60% of the catch within the 'pond' as a whole was made inside the UK's 200 mile fishery limits and around 45% within 50 miles of the British coast.[44]

TABLE 2 : Catch from UK 200 and 50 Mile Zones as a Proportion of the Catch from the EEC Pond (1974) (in '000 tonnes)

Species	EEC 200 miles	UK 200 Miles		UK 50 Miles	
	Total Catch	Catch	% of EEC	Catch	% of EEC
Demersal	1,518.3	752.6	50	529.4	35
Pelagic	1,590.8	1,025.4	65	909.4	57
Industrial*	796.8	522.0	66	282.3	35
All Species	3,905.9	2,300.0	59	1,721.1	44

* Industrial species are those not caught for human consumption but for conversion into fishmeal.

French objections to such attempts by Britain to claim the lion's share of the EEC 'pond' were in part a product of the nature of EEC policy on fishing to which we shall return in the next chapter. However, they were also concerned to underline the economic dependence of particular ports and regions on the maintenance of historic rights around Britain. It was calculated that 90% (by weight) of the catch of Brittany came from the Anglo-Irish zone and that Boulogne depended for 66% of its landings on continued access to that zone.[45] As Figure 4

FIGURE 4 - The location of the grounds fished by Breton trawlers

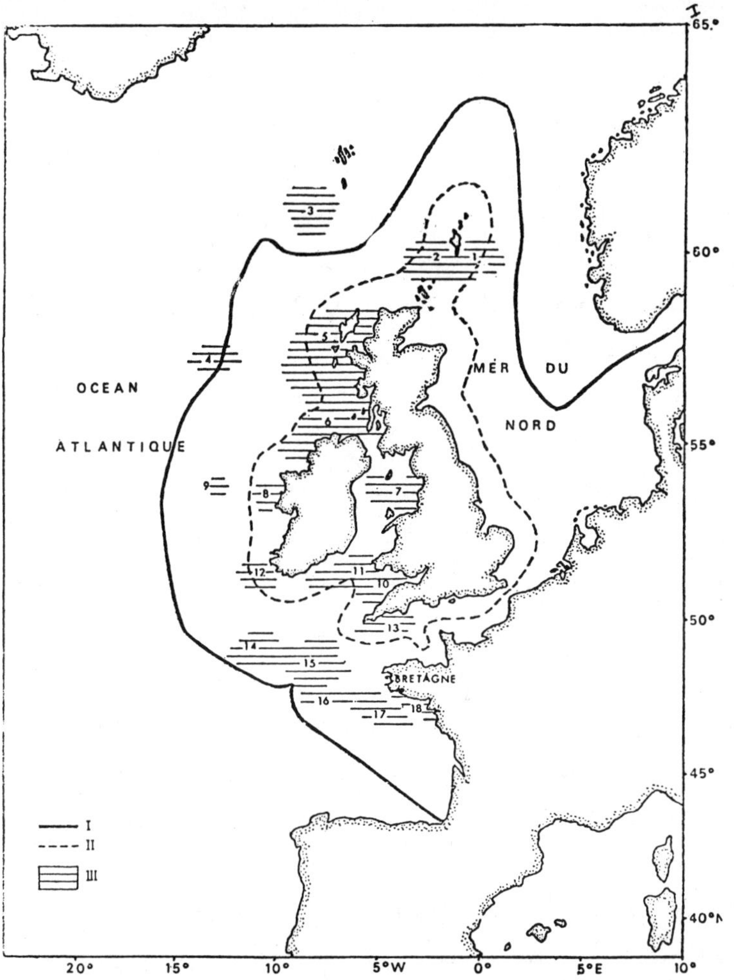

I. Limit of the 200 mile community zone
II. Limit of 50 mile Irish and British zone
III. Areas fished by Breton trawlers

indicates, Breton trawlers were (as they remain) heavily reliant on being able to fish not simply within 200 miles of British and Irish coasts: 10 of their main catching areas were within 50 miles of those coasts.[46] The fear that these grounds might be closed was therefore a potent force inside the French fishing industry.

It would, however, be mistaken to suppose that French fishermen were only concerned to maintain access. In their relationship with Spain, which grew increasingly tense as the period of this study drew to a close, their own sentiments were a mirror image of those of their British counterparts. The prospect of Spanish membership of the EEC made the French industry nervous as to the effects of allowing that country to fish freely within their own waters. In 1977, Spain's catch in the Bay of Biscay at nearly 400,000t. was more than double that of France and French fishermen feared that the large Spanish fleet would devastate the stock levels in that area if they were allowed to enjoy the kind of access that French vessels had around Britain.[47] However, the fact that Spain was still not an EEC country made it possible for this threat to be warded off in the context of agreements beween the Community and Madrid, which limited the levels and areas of Spanish catches.[48] Britain and France's shared membership of the EEC made their own arguments about conservation and access an internal Community matter, and therefore more difficult to resolve.

It should not be supposed that British fishermen were only concerned about the access available to foreign vessels, like the

French. The creation of EEZs by countries like Iceland combined with the increasing pressure on certain stocks, notably herring, to oblige some British fishermen to redirect their efforts around the British coast. This proved a particular problem around the South West of England where large Scottish and East coast vessels came into conflict with smaller local fishermen, who blamed the 'nomads' for ruining their own livelihoods both by reducing their catch and damaging their static gear, such as lobster pots. Thus changes in catching areas generated conflicts of interests within states as well as between them.

4.2 Inshore and deep-sea fishing

It has already been pointed out that the fishing industry is not monolithic but composed of a whole series of interests both upstream and downstream from the catchers. However, the catching sector in both countries is itself not uniform but divided into two important categories, which enjoyed contrasting fortunes in the context of the changes of the 1970s. The basic distinction is between company-owned and skipper-owned vessels: in Britain the former are referred to as the deep-sea industry and the latter as the inshore industry, whereas in France the equivalent terms are 'pêche industrielle' and 'pêche artisanale'.

Although the terms will be retained in this study, they are in both cases not altogether helpful in clarifying the distinction. Thus though British deep-sea vessels have tended to operate in the distant waters of other states, this is by no means universally the case and it is equally misleading to talk of an inshore industry, when its vessels can range long distances from their home ports. As for the French usage, 'industrial fishing' is precisely what everyone in France does not want to do. It refers to the catching of species for reduction into fishmeal which is the chief occupation of Danish fishermen and one of the shared dislikes of British and French fishermen. Both argue that industrial fishing necessarily harms them by producing a substantial 'bycatch' of those fish used for human consumption. The idea of 'artisanal' fishing, on the other hand, is more accurate than the other terms in that in French it does convey the notion that the fisherman or producer does himself own the means of production.

Although it is not universally true in Britain (some boats are managed and owned by agencies),[49] inshore fishing boats and those of the 'pêche artisanale' in France generally belong exclusively or in partnership to the skipper of the boat. Moreover, the proceeds from the catch are divided amongst the crew according to fixed percentages after deducting the shared costs of fuel, ice, food, etc. : the earnings of all on board therefore vary depending on the value of the catch. In contrast the crews and captains of deep-sea vessels are guaranteed a set amount, a variable bonus being paid depending on the success of the voyage. All on board are therefore employees, hired and fired in accordance with the needs of the shore-based companies which own and run the vessels.

The basic criterion for distinguishing between inshore and deep-sea fishing relates to the character of the vessels involved.

Nevertheless, it remains true that despite some overlap, they are also distinguished by size and the range of their operation. Thus the British deep-sea fleet has traditionally been divided into three categories[50]: the distant water trawlers of over 140 feet with their fishing concentrated within the 200 miles of third states, such as Norway, Iceland and the Soviet Union; middle water trawlers of between 110-140 feet, fishing off the Faroes and the North and West of Scotland and near-water trawlers of under 110 feet, whose fishing is pursued in the North, Irish and Celtic Seas. The term 'inshore' refers generally in Britain to vessels of below 80 feet wich operate within the UK 200 mile zone, but the range is enormous: skippers have bought ships up to 135 feet in length,[51] operating for long periods as far away as Norway and the Faroes, while many boats of under 40 feet are at sea for less than a day.

The French, for their part, make a fourfold division of the fleet to underline differences in terms of size and range of operation.[52] 'Petite pêche' refers to vessels at sea for less than 24 hours and, 'pêche côtière' to trips of between 24 and 96 hours: in both cases the vessels are under 50t. in weight and belong exclusively to the inshore or artisanal category. Then comes 'pêche au large' which applies to ships weighing up to 1000t. which are at sea for between 96 hours and 20 days and keep fish fresh on ice at 0°C: within this category the majority comes from the deep-sea or industrial sector but there is a degree of overlap with cooperatives and even some individuals operating ships of this size. Finally, there is 'grande pêche' where the vessels, generally over 1000t., are away for more than 20 days and the fish has to be frozen at up to -40°C: this freezer fleet, exclusively deep-sea in character, is itself divided between those operating off Canada to catch cod, those sailing off the West coast of Africa for tuna and those looking for lobster off South Africa and Madagascar.

Despite a shared division between inshore and deep-sea in France and Britain, the structural balance between the two sectors in the two countries was, until the end of the 1970s, rather different. In Britain, the deep-sea sector was the larger one with roughly two thirds (c.14,500) of the country's fishermen working in it in 1975 and just one third (c.7,500) in the inshore sector.[53] By contrast, in the same year, there were over 25,000 fishermen in the 'artisanal' sector in France and only 6,000 in the deep-sea industry.[54] Moreover, within the British deep-sea sector the nature of the companies involved was quite different from those in France: they had a far larger number of trawlers and a wider range of ports from which they sailed. Back in 1970, for example, Associated Fisheries had 79 trawlers, Boston Deep Sea Fisheries 61 and Ross 52 and they were distributed around the British coast at Grimsby, Hull and Lowestoft in the East, Milford Haven and Fleetwood in the West and Aberdeen and Granton in Scotland.[55] The French, for their part, were more concentrated in single ports with each company owning a much smaller number of vessels: the largest firm in Brittany, Jègo Queré, based in Lorient, never had more than 20 ships.[56]

The impact of change upon this structure was felt much more keenly by the deep-sea companies. Their larger, more expensive vessels, many of which had only been completed in the early 1970s, never had a chance to make money in the changed environment. Several companies in both countries either went out of business or drastically reduced the size of their fleets. This can be seen by observing the way in which the overall shape of the industry evolved. In Britain in 1974 there were 6,461 vessels under 80 feet and 455 over 80 feet: seven years later in 1981, the former category had expanded to 7,106, while the latter had declined to 245.[57] Over the same period in France, the artisanal fleet did decline by nearly 17% from 12,794 to 10,656 boats but the predominantly industrial 'pêche au large' and 'grande pêche' collapsed still more spectacularly in size by nearly 55% from 474 to 217.[58]

Two reasons for this collapse can be identified. First the creation of 200 mile EEZs had a more severe impact in both countries on the deep-sea industry, which depended to an important degree on access to the waters of third countries, than it had on inshore vessels directed to less distant waters. However, this was by no means the only problem. In Brittany, for example, despite continued access to British and Irish waters, upon which the deep-sea trawlers depended for 85% of their catch (as compared with 35% for the inshore vessels)[59], there was still an important decline in the number of these vessels in Breton ports: at Concarneau, for example, the number fell from 107 in 1970 to 48 at the beginning of 1981.[60]

The second and perhaps more important difficulty for the deep-sea industry in both countries was that it found itself squeezed between a higher rate of increase in costs and a lower rate of increase in fish earnings than those found in the inshore industry. The White Fish Authority (WFA) in Britain calculated the costs and earnings per vessel in each of the length categories for the period 1973 to 1977 and concluded:

> "Only in the case of inshore vessels did profitability improve generally over the period. The near water fleet and the freezers (over 140 feet) were unprofitable in four of the last five years and the middle water vessels are only barely profitable. In none of the deep sea groups, in particular, would profitability be sufficient to cover the cost of new vessels."61

The situation was similar in France with the bulk of the deep-sea industry unable to make money, given the balance of costs as against prices.

By contrast, the inshore fleets proved better able to cope. This was partly because of the kind of fish they caught. Deep-sea trawlers in both countries concentrated all their efforts on the so-called 'espèces communs' or common species; many inshore vessels, particularly in France, were more interested in the 'noble' species or 'espèces fins'. The importance of this difference is that 'common' species, such as cod and haddock are very much more susceptible to competition from frozen imports than are the fresh, 'noble' species, such as Norway lobster or sole. One effect of this variable degree of integration into the international market was that prices of the 'noble' species proved more resilient. The price difference can be verified in the French case by observing how the average prices of fish changed in the decade from 1971 to 1981 in the three top ports, Boulogne, Lorient and Concarneau.[62] At Boulogne, a predominantly deep-sea port, the average price per kilo rose from 1.45FF in 1971 to 3.67FF in 1981; at Lorient and Concarneau, where the 'artisans' play a much more important role, the rises were respectively from 2.17FF to 6.61FF and 2.5FF to 7.14FF per kilo. Thus not only were prices at a higher level in absolute terms in each year in the two Breton ports but the rate of increase between the two years was also higher: just above and just under 300% in the case of Lorient and Concarneau, 250% in the case of Boulogne. In ports like Le Guilvinec and Sables d' Olonne, exclusively inshore in character, average prices had improved even more reaching 9.92FF and 11.87FF per kilo in 1981.

However, it was not just the mix of species caught that helped the inshore fleets. In Britain, in particular, inshore vessels were by no means invariably interested in valuable species: it was true of South West fishermen but not of most of those in Scotland, who were eager to go after the 'common' species, such as cod and haddock. What did bind together all inshoremen in both countries was their attitude towards work. The author was often told in Britain that deep-sea fishermen needed incentives to go to sea: the more such incentives were lowered by reducing the scope for bonuses and decasualising the work, the less likely it was that men would be prepared to take the undoubted risks involved.[63] Certainly it is clear that the problem of incentives was less severe in the inshore industry where payment is by results and where there is far less of a gap between the vessel owner and the rest of the crew. This situation creates a set of shared values where:

> "the 'risk' involved in fishing is effectively shared between capital and labour, since the rewards of each are purely a function of the fish that is landed and sold."[64]

The members of the crew can see the possibility of themselves moving up to become boat-owning captains, as they gain in experience and save up. The result is that everyone on board works at rates which would be quite unacceptable in other walks of life and are indeed strongly resisted in the deep-sea industry. In both countries such an outlook protected the inshore industry in a way unmatched in the company owned vessels.

However, the difficulties of the deep-sea industry were not universal. The case of the French tuna freezer vessels of the 'grande pêche' provide an important exception to the general picture presented: the particular conditions applicable to it themselves help to explain why this was so. To begin with, the owners were able to guarantee a working environment which was (and remains) the envy of the rest of the French fishing industry. High earnings (6,000FF per month), long holidays (4 months on, 2 months off), less risk (no fishing at night, for example) and more comfort (the boats are very large and very spacious) earn the 500 French fishermen involved the title of the industry's aristocrats and contrast dramatically with the condition of those working in cold, northern waters.[65]

Like the rest of the deep-sea industry in France, the tuna fleet was threatened by the changes of the middle 1970s.[66] The drop in meat prices in the United States decreased the consumption of tuna and pushed the excess supply onto European markets, where prices fell as a result. However, the market recovered and after that, with demand outstripping supply, the level of prices rose steadily. At the same time, the companies continued to be able to make increased catches with roughly the same number of ships: in 1975 33 vessels landed approximately 48,000t. while in 1981 31 landed nearly 75,000t. In part this was because despite competition from other fleets, in particular the Spanish, there was not a problem of overfishing as there was in the North East Atlantic. The increasing sophistication of new vessels, many operating with helicopters, enabled the captains to take advantage of this favourable supply situation. In part, too, the prosperity of tuna fishing reflected the fact that whatever the theoretical impact of 200 mile EEZs, most West African states were not in a strong position to enforce their rights but were prepared to make a deal permitting the French ships to continue operating. As for fuel costs, there were less difficulties than might be supposed because the vessels operated from Dakar and Abidjan, returning to France only every third or fourth year. Moreover, from 1979 new vessels were equipped to use heavy fuel oil costing 30% less than normal diesel fuel.

The example of the tuna fleet illustrates that under certain conditions the deep-sea industry could avoid decline. These conditions were, however, not solely economic in character. In Chapter Five we shall return to the way in which the relationships between the industry and the government in the two countries also affected the way in which decline was managed. In the French case full advantage was taken of a favourable conjuncture of conditions; in the British case, the conditions were less favourable but little attempt was made to improve them.

4.3 The mix of species

The differential impact of change upon the two industries needs also to be set in the context of the different British and French fish consumption patterns. Traditionally, the British market has been dominated by demersal white fish, notably cod in England and Wales and haddock in Scotland, with a lesser role for the pelagic herring: thus in 1975, these three species alone accounted for nearly 55% by weight of all landings by UK vessels in the UK, 467,000t. out of a total of 869,000t.[67] France, by contrast, has always had a much more even spread of pelagic and demersal species, with shellfish (oysters and mussels) contributing the dominant share of the catch. In 1976, for example, no fish of the water column and seabed exceeded 11% of the total fresh fish catch by weight, whereas oysters and mussels made up over 22% of that same catch.[68]

The years that followed were ones that were particularly difficult for the British to cope with and less difficult for the French market because of these different degrees of concentration. Cod landings, which were taken predominantly from the colder northern waters, inevitably declined in the late 1970s as access to these waters was reduced or eliminated, while over-exploitation of stocks led to a complete ban on herring fishing from 1978 to 1981. Ironically, the overall British catch actually rose in 1978 to over 950,000t.[69] But this increase was the result both of the deep-sea vessels, which were largely deprived of the opportunity of fishing for cod, and the herring fleet, unable to continue as before, turning their attention to mackerel stocks. This fish started to appear in unusually large quantities in the early autumn off the North West of Scotland and to migrate later in the year to the South West of England: the fishermen followed it. In 1978 the mackerel catch shot up to 33% of all landings, while the combined demersal catch had fallen by nearly 100,000t. since 1975.[70] Though the latter catch remained steady over the following years, by 1982 the marked decline in the former catch brought with it an equivalent drop in the overall level of landings to under 800,000t.[71]

What made these developments still more difficult for the fishermen concerned was that the mackerel catch was of very low value and was being caught in quantities far greater than the domestic market could assimilate. In 1978, for example, mackerel was fetching less than £100 per tonne as against the average of over £400 per tonne for the demersal species.[72] To get rid of the fish, the fishermen transshipped it to Eastern European factory ships (entitled 'klondykers' because of the way they gathered in the catch from the British boats). They had been deprived of the right to fish themselves in British waters after 1977 but had a captive domestic market to which they could now supply mackerel. The British consumer, by contrast, was not obliged to eat this fish, despite the drop in the demersal catch: white fish imports increased so that demand could be met. By 1979 they had reached 277,000t. as compared with 173,000t. in 1976.[73]

Whereas in Britain domestic landings of the most commonly eaten species declined and were replaced by frozen imports of the same species, in France it proved possible to keep up the level and balance of national production of the fresh species preferred by the consumer. In 1982 shellfish still represented the most important category of catch at just over 190,000t.[74] Amongst other species there remained no dominant contributor to total production: of the non-frozen catch, only one fish, saithe, contributed more than 10% of the catch, with seven other species contributing between 4% and 10%. The low value mackerel, for example, was one of these seven, but landings did not rise in the same dramatic way as they had done in Britain. Catches of higher value cod and haddock, remained greater in quantity, even though they, too, were nothing like as large as equivalent British catches.[75]

Overall, therefore, the contribution of the fresh fish sector changed little during the 1970s: it continued to represent 85% of

French production in 1979. This is not to say that patterns of consumption remained the same. 42% more in terms of processed products were consumed in 1979 than in 1970 with deep frozen fish up 114%. What's more the French catch contribution to the latter declined from 53% to 35% by weight: imports were up from 32,000 to 89,000t. Despite the resilience of the fresh fish sector, there was still a growing fear that it might find increasing difficulties in selling its catch as the food companies increased their efforts to penetrate the French market.[76] However, the problem was a rather different one from that facing the British, where imports had already taken the place of fish which had once been caught by UK vessels. Both countries were certainly faced with the difficulty of devising a response to imports but the mix between species had helped to make it a more intractable one for Britain.

4.4 The balance between ports

In both countries the association of particular ports with particular kinds of fishing had an effect on the distribution of the costs of change between regions. Once again it was Britain that witnessed the greater upheaval but France, too, was not untouched, as the two countries witnessed the need for adjustment to the decline of the deep-sea industry.

The fate of Hull has already been cited in this study (p 40 above) but it is worth recording the way in which Hull slipped down the table of landings by value between 1976 and 1982.

Table 3 Top five British ports by value of UK vessel landings (£m) in 1976, 1978 and 1982.[77]

1976		1978		1982	
Hull	28.7	Peterhead	29.4	Peterhead	43.7
Grimsby	26.6	Grimsby	28.2	Ullapool	24.9
Aberdeen	16.9	Aberdeen	26.2	Grimsby	21.8
Petershead	9.8	Hull	17.9	Aberdeen	21.2
Mallaig	8.0	Lowestoft	13.9	Fraserburgh	8.9

In 1976 Hull was still the premier British port by the value of her landings but six years later when the catch was down to just over 15,000t. she no longer figured in the leading group of ports by value. Her Humberside rival was still in the group, even if with a reduced catch, but this was thanks to a fleet of 200 inshore vessels, concentrating their efforts on the North Sea. This partially made up for the decline of the big trawlers.[78] Hull, by contrast, had never had an inshore tradition. At least 80% of her catch had always come from Iceland, the Barents Sea and the Norwegian coast and she suffered accordingly when these waters were either closed or subject to strict quotas and licences.[79] Not that her fate was unique - Fleetwood which depended predominantly on hake from Iceland suffered very badly too - but Hull was unique in the scale of its decline.

However, what Table 3 also shows is the way in which the centre of gravity of the British industry moved from England to

Scotland, and in particular to the Scottish inshore fleet, concentrated north of Aberdeen. In the late 1970s, 'artisanal' Scots were already taking 50% of the total UK catch by weight but only 40% by value; in 1980 they had overtaken the fleets in the rest of Britain in value terms as well, catching 60% by weight and 55% by value of the UK catch.[80] They had become the new centre of a reduced industry.

The situation in France was subject to much less turmoil. Boulogne remained easily the first port of France in terms of the volume of fish landed (over 115,000 t. as against Lorient the second port's 70,000 t.).[81] However, Table 4 shows that the changing relative value of species was to the disadvantage of Boulogne which ceased to be the most important port by catch value.[82] By contrast the strictly inshore ports, like Le Guilvinec and Sables d'Olonne, saw the size and value of their catch both rise. However, no ports experienced as dramatic a change in fortune as Hull or Fleetwood. In 1982 La Rochelle was still catching over 100 million FF of fish, even though its catch was under 10,000t.and Lorient's situation looked reasonably healthy despite contraction in the industrial fleet from 42 to 25 trawlers between 1975 and 1982.[83]

TABLE 4: Top five French ports by value of landings (million FF) in 1976 and 1982

1976		1982	
Boulogne	273.0	Lorient	430.6
Lorient	267.7	Boulogne	426.0
Concarneau	156.6	Concarneau	411.0
La Rochelle	114.1	Le Guilvinec	174.6
Douarnenez	64,6	Sables d'Olonne	122.5

The French industry continued to gravitate around the twin poles of Boulogne and Brittany. As Figure 5 indicates,[84] in 1976, when France was divided into four 'Directions' for administrative and statistical purposes, the Nantes Directorate's area was by far the most important with just under half the total catch by weight and just under half the total number of ships but with more than half the catch by value and considerably more than half the tonnage of the French fishing fleet. However, the figures rather exaggerate the gap between Brittany and the Boulogne area. By 1982, a sixfold division was introduced with the Nantes Directorate divided into three. As Table 5 indicates, the result showed a much clearer balance between the ports of Brittany and those of the North. What is more, it illustrates the even spread of fishing activity in France in contrast to the situation in Britain only ten years before when Hull's landings of over 230,000t. per year amounted to more than the total of any of the six regions of France in 1982 and almost as much as the four smallest regions put together.

FIGURE 5 - The relative importance of the fishing areas of France (1976)

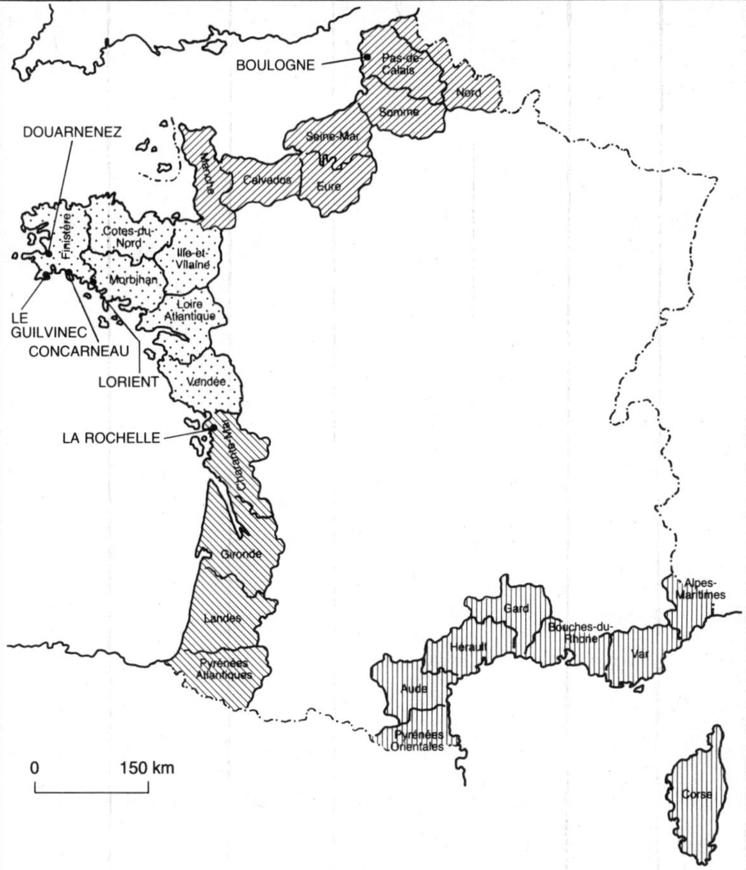

Direction	Fish caught* by weight (tonnes)	Fish caught* by value (FF)	Ships no.	Ships tonnage
Le Havre	170,684	490,648	1,406	52,798
Nantes	201,936	827,145	5,793	150,653
Bordeaux	22,481	164,889	2,346	41,377
Marseilles	30,036	166,979	3,219	16,248
TOTAL	425,144	1,649,661	12,764	261,076

*excludes shellfish

Source: Statistique de la Marine Marchande (1977)

TABLE 5: Catch by volume and value in 1982 in the 6 French regions.[85]

	Volume (tonnes)	Value (millions FF)
North-Normandy	224.932	1,200
Brittany North	81,156	353
Brittany South	168,778	1,592
Loire Atlantique/Vendée	42,573	401
South West	78,412	568
Mediterranean	53,318	533

Thus in terms of where they fished from, as well as in terms of what they fished, who fished and where, the two industries display distinct structural characteristics which were affected in different ways by the changes of the late 1970s and early 1980s.

5. Conclusion

The importance of the discussion in this chapter for the study as a whole is threefold: it underlines the changing relationship between the catching sector and other sectors of the fishing industry; it illustrates the conflicts of interest within the catching sector, both inside and across national boundaries; and it points to the kinds of questions that governments in both countries were obliged to answer in managing economic change within the industry.

First of all, fishermen emerged from economic isolation. Though they had always been linked to a whole network of other economic agents from the boat-builder to the fish shop via the fish merchant, the change in economic conditions underlined the interdependence between them. What happened to one was likely to have an effect on the others. In Britain the failure of the deep-sea fleet to continue to provide regular supplies had the effect of cutting by half to 3,000 the number of fishmongers' shops and giving a dramatic boost to the fish van trade.[86] In both countries, shipyards witnessed a major break in the number of orders for vessels after 1975, which put many out of business and encouraged others to look abroad for business.[87]

Fishermen, too, became more aware of the implications of their relationship with those closest to them in the economic chain. The turbulence of the market underlined that major differences in reward might come from the same amount of effort. Hence fishermen looked more critically at the activities of dockers, merchants and processors. As early as 1971 the bulk of the inshore vessels at Aberdeen moved to Petershead because they no longer wanted to have to pay for the unloading of their fish by so-called 'lumpers' but to do it themselves. At Lorient also, unloading charges were a major issue with inshore boats regularly threatening to go elsewhere, because they lost 15% of the value of their catch in labour charges.[88]

As for the merchants, it became much clearer that they were competing with fishermen in the market for the margin between what the consumer would pay and what the trip to sea had cost and that they were not dependent for supplies on the fishermen of a particular region or even country. Hence, the regular pressure from fishermen for import controls, despite the fact that the processors, the major customers of the merchants, were themselves constantly closing down factories. In France, the number of processing plants was down from 175 in 1964 to 52 in 1978, and their future was unlikely to be improved by measures taken to protect the catching sector.[89] The point was put more colourfully in 1976 by one of the Grimsby buyers when the British Fishing Federation (BFF) suspected their loyalty in accepting Icelandic fish:

> "They (i.e the BFF) are operating the most decrepit trawlers in Europe and if there is a ban on Icelandic trawlers coming down, then those trawlermen who are coming home to the dole queue won't get near the labour exchange for thousands of redundant lumpers, process workers, drivers, friers and merchants."[90]

However, it was not just a question of vulnerabilities between those inside and outside the catching sector. The second feature of the changing situation was the growth of vulnerabilities within the catching sector itself. In the deep-sea industry, the higher level of costs put increasing pressure on the owners to keep labour charges at a minimum and this brought them into conflict with their crews who resisted the attempt to impose the burden of change upon them. Even in the inshore industry escalating costs tempted some to tamper with the percentages used to calculate how much all on board should earn.[91] In both sectors there was an increased awareness of the impact of the activities of one group of fishermen upon another. This might be a conflict between fishermen of the same nationality, for example, pitting the South West boats against those from Humberside and Scotland on the mackerel grounds. Or it might be a dispute between fishermen of different nationality, be it British against French over access or the two of them against the Danes in the dispute over industrial fishing. The conflicts themselves were not new: there had always been disagreements between fishermen of one or more nationalities. What was new was their frequency, intensity and salience.

Finally, this chapter raises questions as to the role that the respective governments could and should play. Given increased interdependence between the various sectors of the industry, how should state institutions respond? In particular, should they see it as their role to balance the interests involved or to impose a particular conception of how the interests in the industry should be interrelated? Given the divisions within the catching sector, how far could those institutions go in encouraging unity, how far should unity be seen as the task of the industry itself to achieve? And given the importance of the increasing integration of fishing into the international economy, how should the British and French states define their roles in defending a domestic industry against pressures from beyond their borders? To begin to answer these questions the next chapter will look not only at the way in which the fisheries issue assumed a different economic shape, but also at how it was thrown into a new political arena, whose character played an important part in channelling British and French responses to the impact of economic change.

3 The political response to change

1. Introduction

One of the lessons of the previous chapter was that the 1970s witnessed an increasing integration of the fisheries sector into the international economy. The development of trade in fishery products, helped by the spread of freezing techniques and by the acceptance of EEZs for exclusive exploitation by coastal states, started to undermine the economic isolation of the fishing industry in Western Europe. Given the different structures of the industries in Britain and France, the impact was not the same in both but equally neither escaped the new pressures, and neither government was able to avoid devising some response to them.

This chapter will suggest that the increasing internationalisation of the economics of fisheries was accompanied by pressures for an international political response. In the process a tension developed between the desire of the Western European states for autonomy and the need for closer cooperation between governments to cope effectively with the issue. The argument will be developed in three stages: first, that the old institutional framework, which contained debate on the fisheries issue at the international level in the 1960s, proved increasingly inadequate as the character of the issue changed; second, that the new framework, provided by the EEC, established a form of institutional interdependence which challenged the autonomy of states to determine the frontier between the foreign and domestic aspects of their fishing policy; and third, that despite the common constraints that the EEC framework imposed, Britain and France still retained different conceptions of the policy implications of the revised character of the fisheries issue.

2. The inadequacy of the old framework

Before the 1970s Britain and France could treat the fisheries issue as a limited sectoral problem. This is not to say that considerable domestic efforts were not devoted to protecting the industry when gluts of fish appeared and prices tumbled. Nor is it to suggest that at the international level, there was no recognition of the interrelationship between national fishing fleets. But in that period it was possible to make a clear institutional distinction between the domestic and foreign aspects of fishing policy: the domestic aspect was one to be dealt with in discussion between the ministry responsible and the representatives of the industry; the foreign aspect was for bilateral negotiation with third countries or for multilateral discussion within the various international fishery commissions, such as NEAFC. Between the two environments the government could act as a kind of 'gatekeeper', protecting the interests of the industry at home in the foreign arena and implementing the results of international negotiation in the domestic arena.

This institutional separation was bolstered by the generally accepted status of fish as a collective good, available to all with very little restriction. The states in the North East Atlantic area broadly agreed in the 1960s not to seek to impose domestic costs on one another by limiting access to their waters, on the assumption that all could continue to make increasing catches under an 'open seas' regime. They recognised the need for a degree of international cooperation but only to the extent that it did not impinge seriously on domestic policy. Hence there was no serious challenge to the legitimacy of the governmental 'gatekeeper' role.

The fisheries commissions were the institutional expression of such governmental attitudes to the international arena. The most important from the point of view of this study was NEAFC. It was established to supervise the North East Atlantic Fisheries Convention, signed in 1959, within the area indicated in Figure 1 (see p 47 above) and continued to represent the chief forum of international collaboration on fisheries until the EEZ doctrine became a reality in the area of its jurisdiction in 1977. How little governments were prepared to be restricted in the exercise of their sovereignty by this institution can be seen by examining the nature of its operation and powers.

Firstly, it was a body of very limited resources: for a long time its staff amounted to three people, the Secretary and two typists. As for the Secretary, he "had always been a British civil servant who spent three quarters of his time working in the British Ministry of Agriculture."[1] Though the delegates of the member states to the Commission (two for each of the 16 members) had the right to appoint such staff as they chose, they were not concerned to create a secretariat which might challenge their own right to manage fisheries.

Secondly, the meetings of the Commission were held in a political vacuum. The Commissioners from each country were civil servants - in the British case, the Fishery Secretaries from the Ministry of Agriculture, Fisheries and Food (MAFF) and the Department of Agriculture and Fisheries for Scotland (DAFS) so that outside the trade-related journals and newspapers there was next to no debate as to what was at stake. The containment of discussion in this way also meant that there was no possibility of generating political deals between this issue and other issue areas.[2] Fishing remained strictly separate and even within the limits of the fisheries sector, the narrowness of the Commission's duties excluded any 'linkage politics'. Thus the Commission was not concerned with the extent of fishing limits, the prices obtained for fish, the aids given by governments to their own industries or the negotiation of agreements with non-member countries: these were either matters for individual countries to undertake themselves or as in the case of limits, matters to be settled by inter-state negotiation, such as that which had led to the London Convention of 1964.

What, above all, restricted challenges to the sovereignty of member states, was the limited character of the Commission's powers. Its major task was "to consider...what measures may be required for the conservation of the fish stocks and for the rational exploitation of the fisheries in the area."[3] And yet the Commission's powers to carry out this task were only ones of recommendation. A straightforward procedure existed to enable a state to object to any recommendation and thereby escape the need to implement it. Indeed once three states had objected, all the contracting parties were relieved of the obligation to heed the recommendation.[4]

The arrangements did not, however, pose difficulties as long as the 'open seas' regime continued to operate and everybody's catches continued to rise. What started to undermine them was the increasing awareness of the importance of taking conservation seriously as stock levels began to drop at the beginning of the 1970s. At that point the effects of the voluntary regime were clearly exposed. A system of mutual inspection, known as the Joint Enforcement Scheme, was introduced and it revealed that a number of states had been less than enthusiastic in enforcing regulations against their own nationals in earlier years. The percentage of inspections resulting in reports of violations of regulations against vessels of third states, compared with the percentage when those states were themselves responsible for inspection, rose considerably: in the French case from 6.7 to 44.7; in the British case from 3.6 to 24.5. Even with the scheme, however, prosecution remained the prerogative of the port state and not the coastal state in whose waters the offence was committed.[5]

In 1974 a further move was taken to improve conservation, when it was agreed to introduce quotas known as Total Allowable Catches (TACs). Here, it seemed, was a way of getting round the difficulty of enforcing regulations on the mesh sizes of nets,the traditional mechanism of control. But it proved impossible to establish levels of catch that would not mean significant hardship for any of the member states. Whatever the recommendations of the non-governmental experts of the International Council for the Exploration of the Seas (ICES), the Commissioners would regularly increase the permitted catch so as to ensure agreement. Table 6, presented to the Trade and Industry Subcommittee, underlines the discrepancy between the two for 1976 and allows a comparison with what was declared as caught.[6] Of the 31 TACs recommended by ICES, the Commission increased the figure in 14 cases and in 6 of these 14 cases, the actual catch exceeded even what the Commission proposed. In 16 cases where the Commission proposed no TAC on the basis of the ICES recommendation, the declared catch was higher than the recommended one on 11 occasions, often by very substantial amounts. The whole system was described as "the madness of NEAFC" by the editor of Fishing News who commented before the decisions for 1976 were taken:

> "Here we have supposedly responsible and intelligent delegates being advised by their scientists that herring stocks in the North Sea are on their last legs - yet it is almost certain that they will increase the amount of herring some countries can take to get a working agreement."[7]

And what is more, there was widespread recognition that what states admitted they had caught was not what they actually caught. As one observer put it, if the TAC was 50,000t, the state would acknowledge a catch of 100,000t.but everyone knew the figure was really 150,000t.[8]

TABLE 6 - International catches and TACs recommended by ICES for fishe in the North East Atlantic (in '000 tonnes)

Fishery	*Nominal catch* 1975	*Recommended TAC* 1976	*NEAFC TAC* 1976	*Nominal catch* 1976[1]	*Recommended TAC* 1977	*Recommended TAC* 1978
North-East Arctic						
Cod	829	700-800	810 (850)	859	850	850 } 1,000 combined
Haddock	176	100	–	144	110	150 } 1,000 combined
Saithe	233	190	–	221	200	160 } 1,000 combined
Greenland Halibut	33	–	–	33	–	40
North Sea						
Herring	365	140[2]	160	183	0	0
Sprat	641	650	650.5	617	400[3]	400
IIIa Sprat[4]	106	–	–	63	80	80
Mackerel	318	249	–	297	220	190
Cod	188	130-210	236	209	220	220
Haddock	184	106-155	206.25	206	165	105
Whiting	153	160	189	191	165	160
Saithe	267	200	–	326	210	200
Plaice	108	85	99.9	109	71	95
Sole	18	8	12.5	14	6.7	8
Iceland (Va)						
Saithe	88	75	–	79	60	60
Faroe (Vb)						
Cod	39	28	–	40	32	30[5]
Haddock	21	17	–	25	17	17[6]
Saithe	41	50	–	32	40	40
Sub-Area VI						
Cod	13	14	–	13	19	20
Haddock	64	23	–	59	18	12
Whiting	20	13	–	21	22	17
Saithe	27	30	–	36	20	20
West Scotland (VIa)						
Herring	141	66	136	107	48	53
Sub-area VII						
Cod	20	18.1	–	19	17	20
Haddock	8.6	7.6	–	–	6.5	8
Whiting	32	19.5	–	13[7]	20	25
Irish Sea (VIIa)						
Herring	24.5	–	–	22	12	12.5
Plaice	4.6	4.0	4.15	3.28	4.0	4.0
Sole	1.44	1.6	1.67	1.38	1.4	1.4
West Ireland (VIIb.c)						
Herring	17	–	–	19	10[7]	10[1]
English Channel (VIId.e)						
Plaice	2.82	3.26	3.34	2.57	2.0 (VIId) 0.45 (VIIe)	2.5 (VIId) 0.6 (VIIe)
Sole	1.33	1.36	1.45	1.69	1.00 (VIId) 0.45 (VIIe)	1.15 (VIId) 0.35 (VIIe)
Bristol Channel (VIIf)						
Herring[6]	3.4	–	–	–	1.0	1.0[1]
Plaice	0.47	0.5	0.64	0.3	0.4	0.4
Sole	0.57	0.7	0.7	0.52	0.4	0.6
Sub areas VI, VII and VIII						
Mackerel	492	295	–	465	250	240

From another perspective the results of NEAFC's deliberations were an inevitable consequence of a voluntary form of international cooperation with no direction from any central authority. In such a system national authorities are necessarily very reluctant to relinquish any of their powers and thereby to suggest that they are giving up their role as the 'gatekeeper'. To do so is to suggest not only that the domestic and foreign arenas are no longer clearly separable but that the distinction lacks legitimacy and thereby gives the right for non-national authorities to intrude into the domestic sphere.

However, whatever the behaviour of delegates within NEAFC, the passage of the 1970s revealed that the fisheries issue no longer enjoyed the status it had had before. Fish were still a 'collective good' in that they remained potentially available to all but the nature of this availability had changed.[9] First of all, it ceased to be the case that the fish catch of one state did not subtract from the catch of another: it was clear that there were Maximum Sustainable Yields (MSYs) and that if they were exceeded, fish stocks were in danger of disappearing. Secondly, it became evident that it was possible to exclude states from the benefit of this collective good by altering property titles: the concept of the EEZ gave coastal states the possibility of preventing others from making fish catches in their waters. Thirdly, the conditions of availability of fish were seen to depend on more than fishing rights: the workings of the international market could render the use of national measures of conservation and exclusion largely irrelevant. As a result, as Chapter 2 showed, the industries had become both more vulnerable to the action of others beyond their borders and the avoidance of vulnerability was impossible. Such interdependence necessarily created pressures for more concerted international action in Western Europe.

However, NEAFC could not seriously hope to cope with this new interdependence. As already indicated, the weakness of its powers made the imposition of severe burdens on domestic fishing industries by the Commission, even when stocks were under severe pressure, politically impossible. Equally, it could have no role in the changing of limits, as it was the inheritor and not the formulator of this aspect of the ocean regime. And its limited remit excluded any intervention in the wider forms of economic interdependence to which national fishing industries found themselves subject.

At the same time, the structural character of the industries of Britain and France increased the pressures for some kind of international response. Both depended heavily on stocks fished by other states in the North East Atlantic and needed some mechanism for controlling the level of catches. Both were subject to the effects of the international trade in fish and its accompanying threat of imports. Even in the case of the setting of limits, where Britain's concentration of fishing on waters close to her coast gave her a quite different perspective from that of France, it is by no means clear that she could have avoided an international response. Whatever the general opinions of the industry and its supporters, the importance of the migration of stocks between zones made unilateral action, like that of Iceland, a hazardous prospect.

As one survey of cod indicated (see Table 7)[10], there is a dramatic difference between the percentage of this fish found in British waters when they are three years old as opposed to when they are one year old. A change of fishing patterns could therefore have rendered British catches of her favourite fish vulnerable to a change in fishing habits by her neighbours, a threat to which Iceland was not subject when she took her unilateral position.

TABLE 7 Distribution of cod caught in EEC waters (% by national zone)

	Denmark	Germany	Netherlands	UK
One year olds	20.3	54.1	12.0	13.6
Two year olds	8.4	37.9	15.2	38.4
Three year olds	6.4	6.7	9.7	77.0

The question of the kind of international response that the two countries should devise to cope with the structural characteristics of their industries was effectively answered by the fact that both were members of the EEC. Despite the distaste of many within Britain, not least inside the fishing industry, for the Community and the belief that better arrangements could have been devised outside it,[11] both countries found themselves in a situation where there was no serious alternative to the EEC as an international forum for coping with the new economic interdependence.

Though NEAFC continued to exist in a formal sense after the end of 1976 and though the CFP had emerged in its original form in 1970, it was only in 1977 that the full importance of the EEC in the fisheries arena emerged. At the Hague in October 1976,[12] it was agreed by the foreign ministers of the Community countries that:

a. the fishing limits of member states in the North Atlantic and North Sea be extended jointly to 200 miles from January 1, 1977;

b. the Commission of the EEC be given a mandate to negotiate all future fishing agreements with third countries as to reciprocal fishing rights;

c. the Commission exercise an overall watchdog role over the level of stocks within the 200 mile boundary.

Despite the uncertainties that this decision left, it radically changed the context and content of fishing policy for the member states. This fact was underlined by the contrast between the character of future negotiations with third countries and Britain's final 'Cod War' with Iceland, which was drawing to a close at the end of 1976. That dispute was to be the last example of the classical bilateral resolution of a fisheries dispute with the government taking the lead in the international area, on the basis of a broad domestic consensus. The maintenance of fishing rights off Iceland had been a foreign policy objective, where the government had played a clear 'gatekeeper' role. The objective was abandoned because of pressure from other governments. On the one hand, there was the spread of the legitimacy of the EEZ doctrine, on the other the advice of Britain's NATO partners, concerned to maintain the use of the Keflavik air base in Iceland. From the beginning of 1977 such a style of inter-state policy making became increasingly inappropriate, as the debate over the status of fisheries moved firmly into the EEC area, with its very different set of processes and obligations.

3. A new framework of institutional interdependence

The difference between the old NEAFC forum and that provided by the EEC was enormous. The shift meant a dramatic widening of the range of institutions with an interest and a role in the making of policy on fishing. Instead of a small group of national civil servants, occasionally interrupted by their respective ministers, there was a mass of new actors, many of whom did not owe a specific loyalty to a national government and who were not therefore concerned to maintain the boundary between the foreign and domestic aspects of the fisheries issue.

First of all, member states were confronted with a quite different creature than NEAFC in the European Commission. Here was a body with the specific tasks under the EEC Treaty (Article 155) of initiating policy, influencing the decisions taken by national ministers, implementing those decisions and ensuring that member states abided by them. What made these powers particularly significant was that unlike the old Fisheries Commission, which could only issue recommendations with no binding force, the EEC Commission formed part of a policy process which could produce directives and regulations that had the force of law within all member states. It did not have the final power of decision - that belonged to the Council of Ministers, made up of the national ministers of the member states - but it could certainly influence those decisions.

At the same time it was not without a degree of independent skill and capacity. Rather than rely on a single Secretary and two typists, the Commission decided at the end of 1976 to establish a separate Directorate General (number XIV) with a staff of 50 or so, with a specific responsibility for fisheries. Moreover, this Directorate General was not without access to expertise and advice outside national governments, both through formal and informal channels. An Advisory Committee on Fisheries had been established back in 1973 with a wide range of representatives from the fishing industry in the member states. In 1979 a Scientific and Technical Committee was set up to provide the kind of detailed information that the Commission needed if it was seriously to contest national definitions of fishery problems. These formal channels of communication were supplemented by informal links. The fishing industry gave sufficient weight to the importance of the Commission that it established its own Brussels based, Community-wide pressure group entitled 'Europêche'. This body helped the main representatives of the industry to coordinate their positions but also enabled the Commission to develop a wider understanding of the problems of the industry.

The Commission was therefore organisationally and legally equipped to start to coax governments to make agreements as well as checking that they were abiding by them. During 1976, for example, the Commission proposed that all member countries should enjoy an exclusive zone of 12 miles around their coasts, as a way of offering a compromise between British demands for a much wider zone and the demands of other states, including the French, that there be no such zone at all for the vessels of EEC member states. Despite the failure of the Council to agree to this proposal later in the same year at the Hague (and indeed disagreement within the Commission itself), this proposal did form the basis of the accord which was eventually reached in 1983.[13] At the very least, the Commission had established the terms of the debate, long though the debate turned out to be.

Similarly, the Commission was involved in checking up on the implementation of agreements. At The Hague, there was no agreement on precisely how to conserve stocks of fish. However, Annexe VI to The Hague Resolution stated that each state could adopt as an interim measure and in a form which avoided discrimination, appropriate measures to ensure the protection of resources situated in the fishing zones off their coasts.[14] All states were obliged to consult the Commission before introducing such measures and the Commission had the job of ensuring that there was no discrimination. The United Kingdom in particular, was the subject of a large number of investigations, after other states complained that her conservation measures discriminated against their own fishermen.[15] The Commission was sometimes satisfied that the measure was non-discriminatory but if not, it had the right under Article 169 of the Treaty to bring the state concerned before the Court of Justice, the second important institution in the EEC structure.

The Rome Treaty gives the Court the specific role of ensuring that Community law is observed (Article 164) and imposes an obligation on member states to comply with what the Court decides

(Article 171). Thus the judges ruled that Britain had indeed acted in a discriminatory way by banning both fishing for Norway pout in the so-called 'Pout Box' in the North Sea from October 1978 to March 1979 as well as fishing for herring off Northern Ireland and the Isle of Man. Though the Court itself has no mechanism for guaranteeing enforcement, states are generally unwilling to be seen to be overtly flouting its judgements. Thus in the 'Pout Box' case, Britain's response was to pursue negotiation with the Danish government, against whose industrial fishing the ban was directed, and to reach an agreement which enabled fishing to continue but in areas where the 'by-catch' of species for human consumption would be less. In other words the Court could act as "a catalyst for the political resolution of the conflict."[16]

Both the Commission and the Court have a rather technical flavour to them. However, the third Community institution of importance, the Assembly or Parliament, has a much more overtly political character, especially since its first election by direct universal suffrage in 1979. Though the Treaty only accords it limited "advisory and supervisory" powers (Article 137), it too plays its part in the blurring of the domestic and foreign policy arenas. As with the Commission this proved possible in part through institutional change. Thus the Agriculture Committee set up a fisheries working party of those members of the Parliament, who had a particular interest in or connection with the industry. The members of the committee were able to pick up pieces of information that came from the domestic industries and give them a much wider airing, often designed to embarrass other member states.

A good example of this political role came in March 1980, after ITV's 'World in Action' had transmitted a programme entitled 'Spying for Survival'. In it a Hull trawler skipper visited Boulogne and claimed to have discovered large quantities of herring on the quayside, when there was supposed to be a complete ban on fishing for it.[17] Immediately the Scottish MEP, Provan, set about showing the film in Strasbourg and writing a report to oblige the Commission to tighten up enforcement procedures. Nor were British MEPs the only ones who were active. During 1980 it emerged that Spanish vessels were registering in Britain and flying the British flag.[18] French fishermen alerted their own MEPs and Mrs. Le Roux, a French Communist, invited the Commission to tell her what could be done about it.[19] Thus she and Provan could be seen to be representing fishing interests, the Commission could be obliged to take a position on national behaviour and the national governments could be forced to explain why such practices were permitted.

However, the changes from the old NEAFC regime were not only visible in the role of the Commission, Court and Parliament: they were also seen in the more prominent role of national ministers. The Council, the body given the power by the Treaty to take decisions on the basis of Commission proposals (Article 145) brought the Fisheries Ministers of the member states together regularly from October 1976 onwards. Within the Council, under the so-called Luxembourg compromise of 1966, member states have the right to veto any agreement which they see as contrary to their vital national interests. At the same time, it is important to recognise that the defence of those national interests takes place in a quite different environment from that provided by NEAFC. Ministers descend from meetings to explain how well they have defended their country's corner but they also know that next week or next month will see another meeting and that the pressure for some kind of agreement will increase: they are, if not 'condemned to agree', at least obliged to think seriously of the consequences of failure to do so. This is all the more likely in an environment, where one state can withold something that another wants in a way that was not possible under the NEAFC arrangements.

Such 'linkage politics' was particularly prevalent in the Council of Ministers in the context of the international agreements which the Commission was authorized to negotiate from 1977 onwards. Throughout 1981, for example, Britain blocked a draft agreement with Canada for the period 1981-1986, which would allow Community vessels to catch a certain volume of fish off Canada in return for easier access of Canadian fish products to the Community. Though the agreement was critical for the German fleet, British fishermen saw in the agreement increased imports of cod from Canada which would undercut their own catch. Only through an improved system of protection against imports was it possible for the Community as a whole to sign the agreement and then it was once again endangered by an extraneous element: the Council adopted in March 1982 a ban on imports into the Community of seal pup skins, a measure originally urged by the European

Parliament. Canada bitterly resented the move and threatened to renege on the fishing agreement.[20]

France, too, was aware of the possibilities of 'linkage politics'. She blocked an agreement with Norway, which was of particular importance to British fishermen, until her own desire to maintain access to the Faroes was granted. This in turn had been blocked by Britain, whose fishermen had effectively ceased to fish there and who saw an opportunity to exercise pressure on France to reach a more suitable settlement on the wider question of access to British waters.[21]

However, it was not simply a question of the isolation of particular parts of the fisheries dossier. Other totally different issues could intrude, particularly at the apex of the Community's decision-making process, the thrice yearly meetings of Heads of State and Government in the European Council. To them the Council of Ministers regularly referred matters that it felt unable to resolve itself. In the Spring of 1980, the British government made a determined effort to settle the problem of Britain's budgetary contribution to the Community. At the European Council meeting, Mrs. Thatcher obtained an interim solution to the problem but the other leaders insisted as part of the bargain, on an agreement in principle to the establishment of a revised CFP by the end of the year. Though the UK government denied any link, on May 30 1980, the Council of Ministers did undertake to adopt "the decisions necessary to ensure that a common overall fisheries policy is put into effect at the latest on 1 January 1981."[22] The fact that the deadline was not met does not detract from the fact that pressure for agreement was increased, following Britain's implicit acceptance of a 'package deal'.

In sum, therefore, it was no longer possible for the fisheries issue to remain isolated as it had been under the NEAFC arrangements. A new set of institutions was prepared to challenge the role of governments as 'gatekeepers' between the domestic and foreign arenas and the right to autonomous decision that such a role implied. They delved into the domestic affairs of member states, allowing individuals and groups from one state to look at the behaviour of another without having to pass through governmental channels. Institutional interdependence had replaced sectoral politics as the environment for managing the issue.

4. Britain and France within the EEC: divergent priorities

Just as the economic changes that hit the fishing industry were not felt in precisely the same way by Britain and France, so the two countries responded rather differently to the abandonment of the old interstate arrangements and integration into the political processes of the EEC. Their interests in what an EEC policy could do for them and what they would have to do within such a policy were far from identical. To illustrate the point this section will look at the divergence in priorities of the two countries and the way in which that divergence developed from the inception of a Community policy in 1970 to the agreement on a revised policy in 1983.

4.1 The initial phase

Though the EEC was to become involved in a much wider management role than NEAFC in the fisheries arena, this did not happen right at the beginning of the Commmunity's development. Fish was defined as an agricultural product under Article 38(1) of the Treaty and was thus incorporated within the common market. However, during the 1960s this meant little more than a reduction in the tariffs existing between the original Six for trade in fish products. Only in 1966 did the Commission produce an initial report on a broader fishing policy and two more years passed before it presented any concrete proposals to the Council of Ministers. Even then there was substantial disagreement as to what a CFP should look like.

How long the deadlock would have continued it is impossible to say. What is clear is that the prospect of four new states (Denmark, Ireland, Norway and the United Kingdom) - all of them with important fishing industries - applying to join the Community served as an important external spur to agreement. Though some countries, like Germany and Holland, wished to wait to discuss the matter with the prospective members, the view that prevailed was that expressed by the Dutch Commissioner, Mansholt:

> "the candidate countries have their own fishing policy; to discuss with them, the Community needs its own policy too."[23]

The result was an agreement on the basic principles of the CFP on June 30 1970, the very day before negotiations with the applicant states began.

British policy and attitudes towards the CFP were to be permanently marked by the character and timing of this agreement. Eight years later the British minister responsible for the negotiations on entry, Geoffrey Rippon, was to say that it was "an act of folly" on the part of the Community to confront Britain with the CFP in this way.[24] What made him and others so upset was that the original six had incorporated into Community law the principle of equality of access for all Community vessels into the waters of all Community countries. Thus as Article 2(1) of the relevant regulation put it:

> "The system applied by each Member State in respect of fishing in the maritime waters coming under its sovereignty or within its jurisdiction must not lead to differences in treatment with regard to other Member States. In particular, Member States shall ensure equal conditions of access to and exploitation of the fishing grounds situated in the waters referred to in the previous paragraph for all fishing vessels flying the flag of a Member State and registered in the Community territory."[25]

All of the applicant states were unhappy with this principle because they had much greater fishing interests than the Six: their total catch was three times higher, they had many more inshore fishermen in regionally sensitive areas and the fish stocks in their coastal waters were much larger.[26] It looked to the applicants like a piece of sharp practice on the part of the Six, guaranteeing them access to fish-rich waters, without being obliged to take the interests of the coastal states concerned into account.

As a result, all four of the applicants pressed for revision of the policy. A MAFF press release in September 1971 explained:

> "HMG consider it reasonable to ask the Community to recognise that the fisheries policy will require modification to meet the needs of an enlarged Community."[27]

However, along with the other applicant governments HMG was faced with a fait accompli which could not be modified substantially, if it was to be successful in gaining entry to the Community. There is little doubt that this dilemma was critical in determining the rejection of membership by Norway in a referendum in September 1972.[28] There is also little doubt that the Heath government was determined not to allow the fisheries problem to block the conclusion of the Accession Treaty. Negotiations only began in June 1971 after all the other issues had been resolved and were not concluded until December, when the government had already in October received a parliamentary vote in favour of the terms of entry as presented in a White Paper produced the previous July.[29]

The agreement that was reached on fisheries and incorporated in the Treaty of Accession (Articles 100-103) provided for 'derogations' from the basic principle of equality of access until December 31 1982. Under this arrangement all member states could maintain a six mile fisheries limit around all of their coast, as well as a twelve mile limit in agreed areas, (which in Britain's case covered 95% of her territorial fishing grounds), provided they allowed foreign fishermen with 'historic rights' to continue to exercise them. Though the government claimed a victory and maintained it would have a veto power in 1982, critics pointed out that it was in the nature of a derogation that it would lapse automatically and that the other members would have a veto over any arrangements to replace those agreed for the ten-year period up to 1983.[30] As it transpired the situation changed dramatically before then but the events of 1970 and 1971 were to make the limits issue a critical one for Britain throughout the period of this study.

For France, by contrast, the debate on limits was not one of such dispute. Certainly she pressed for agreement amongst the Six before the start of the enlargement negotiations.[31] But it is much less clear whether the idea of equality of access was central to that pressure. Her access to the fishing zones around Britain was in no way threatened by the existing narrow limits. Moreover, within France there was considerable argument as to whether equality of access might not conflict with the Treaty's provisions for non-discrimination.[32] Under such a regime fishermen operating in the waters of another state would be free from the constraints of that country's laws and therefore would enjoy a commercial advantage in as far as they escaped regulations on social security and conservation measures to which their competitors were subject. There was therefore support for the idea of freedom of establishment, whereby any ship could fish in the waters of any state, provided it registered in that state and accepted the constraints of national law that this implied. In the event, this argument did not prevail but it underlines that equal access was not free from controversy, even amongst the Six, and that it could be challenged by the use of arguments couched in Community language.

However, there is a further reason for supposing that access was not the central concern of France. She had been constantly worried about the effect of competition from other Community countries and was very eager to ensure that her fishing fleet would be guaranteed a degree of protection, particularly after February 1 1971, when the fish trade inside the Community was to be totally liberalised. It was pointed out that the French industry was poorly equipped to stand up to the other European industries, particularly that of Germany. Though the two industries were making roughly similar catches, France needed 40,000 men to do it and Germany only 8,000. What is more, the Germans had a very modern fleet, while the French were desparately in need of finance to renew their ageing vessels.[33]

The result was that she made it a condition of agreeing to a CFP that a system of market support be introduced. The system derived its inspiration from that operating in agriculture, with an interlinked set of withdrawal, guide, and reference prices.[34] It was designed to enable producers to be guaranteed a reasonable income and to be protected from low-priced imports from outside the Community. Under its provisions the Community provides compensation for a certain number of species if the market price falls below the withdrawal price and the fish have to be reduced to fish meal and oil. The withdrawal price itself is calculated with reference to the guide price, a price established on the basis of what fish prices have been over the previous three years. Furthermore, if non-EEC imports fall below the reference price, they can be blocked to prevent loss of earnings by Community fishermen.

In view of the difficulty of running such a scheme centrally, it was decided to adopt a procedure used in some sectors of EEC agricultural policy and to devolve responsibility to regional or local Producer Organisations (POs) within each of the member states. Their role is to manage the day-to-day operation of the price system and more generally, to ensure the orderly marketing of fish products. Again it was France that was most enthusiastic about the establishment of POs. As we shall see in Chapter 8, she already had equivalent institutions operating in the domestic arena and she pressed hard for the POs to be similarly structured and financed. The other states were by no means as keen - they feared an excessive degree of 'dirigisme' - but it was accepted that 50% of the money required to set up a PO would come from the Community, aid to be spread over the first three years of the organisation's existence.[35]

At the same time, France was successful in getting agreement that money should become available from EEC funds to restructure national fishing fleets. Under these arrangements she was able to finance the conversion of her 'grande pêche' fleet of salted cod and tuna ships into a predominantly freezer fleet. All

Community states, including France, were also authorised to grant national financial aids to the extent that they were necessary to achieve the aim of giving EEC fishermen "equal rights to carry out fishing operations."[36]

It is true that even at the time French observers were not unaware of the benefits of the equal access provision. As the President of the CCPM put it,

> "The entry of Britain into the Common Market will alter the status of its waters to our advantage, particularly after the ten year derogation period, when all its waters will become shared."[37]

However, 'equal access' was certainly not the dominant concern of French policy makers. Their main desire was to use the Community to protect French fishermen from the impact of market pressures. Within the United Kingdom, this aspect of EEC policy went largely unnoticed: the question of limits and access was seen as the important one and hence it was too easily considered to have been critical in the thinking of other member states. As a result the development of attitudes and policies were guided by very different logics over the years that followed: France saw the possibilities of gaining further economic support from the Community and pursued them very actively; Britain continued to believe that the policy had been biassed against her from the outset and sought to rectify that bias, adopting an essentially defensive posture.

4.2 The development of policy

Though the regulations of 1970 established a base for Community policy development, important issues remained outside the EEC's sphere of competence. The EEC had no role from 1970 to 1977 in the agreement of conservation measures. These remained the preserve of NEAFC. Similarly, there was no question of the member states losing control of negotiations with third countries over fishing rights. In 1972 France negotiated a bilateral deal with Canada which offered guaranteed access for her cod fleet until 1986 and Britain pursued her conflict with Iceland in 1972-3 and 1975-6 to try to maintain access for her deep-sea fleet, both of them acting on a bilateral basis. Britain did use the Community forum to block a free trade agreement between the EEC and Iceland[38] just as France obtained Community finance to renew the vessels sailing into Canadian waters. However, the Community remained for both a source of opportunities rather than constraints in their relations with third countries.

At the same time, the two states continued to concentrate their attention on the aspects of Community policy which had been of most concern to them at the beginning of the 1970s: Britain on the issue of access, France on the question of markets. However, Britain found itself in a difficult situation because of the divided character of her interests in relation to the extent of fishing limits. The inshore industry was predominantly concerned to seek protection from the intrusion of foreign fishing vessels into British waters, and looked to the example of Norway which established a 50 mile limit after deciding not to join the EEC.

However, the deep-sea industry worked to maintain access to the waters of third states and saw Norway's non-membership as a severe blow to future fishing possibilities. As long as accepted international limits remained narrow, the conflict of interest between the two sections of the industry remained a potential one but from 1973 onwards the increasing acceptance at the UNCLOS III conference of 200 mile EEZs, and the Icelandic extension of limits to 200 miles in October 1975, combined to make the conflict an actual one. If individual countries had the right in international law to 200 mile EEZs, then the British deep-sea fleet had no right to continue to fish off Iceland whereas the inshore fleet could expect a wider zone free of foreign competition.

The result was that Britain appeared to pursue different strategies in the EEC than in the wider international arena. In the latter, given her continuing conflict with Iceland, she could hardly abandon her traditional stance and she did not:

> "Until late in 1976 the UK followed the doctrine of customary international law that coastal states' exclusive rights to fisheries in the seas adjacent to their coasts should be confined to narrow limits, whether within a territorial sea or special fisheries zone, and that in the area of high seas beyond those limits freedom of fishing prevailed for all states."[39]

At the same time, negotiations were continuing in the Community on the proposal of the Commission, presented in February 1976, and eventually agreed at The Hague in October, that there should be a joint extension of limits to 200 miles by EEC member states. That agreement effectively ended any chance of challenging the right of Iceland to claim a 200 mile limit and to exclude British deep-sea trawlers from her waters.

However, at the same time as she was expelled from Icelandic waters, Britain found herself once again in difficulty over limits with her EEC partners because she was not in a position to claim a 200 mile EEZ for herself. As she had agreed to the 'equal access' provision of the Treaty of Accession, subject to the ten-year derogation, it was politically impossible to take the kind of unilateral action that Iceland had taken. This was effectively recognised by the government when as early as May 4 1976, Roy Hattersley, the Minister of State at the Foreign Office called for EEC members to be given exclusive rights in zones varying in extent from 12 to 50 miles.[40] The effect was to reinforce concentration on the limits issue as well as the belief that Britain had been doubly penalised by expulsion from Iceland and membership of the EEC.

The French, by contrast, never faced this same conflict over the question of limits. They quickly recognised the way developments were moving on the international stage. As early as 1972 President Pompidou assured the leaders of the fishing industry that France would fight to ensure that the duties as well as the rights of coastal states would be considered at UNCLOS.[41] The leaders themselves acknowledged that France would have to agree to the establishment of EEZs but insisted that coastal states be obliged to pay attention to 'droits acquis' or historic rights.[42] No effort was expended in fighting the spread of the EEZ idea but increasingly opinion hardened around the importance of maintaining access to British waters. After The Hague agreement Giscard d'Estaing made it quite clear that he favoured "the sharing of the resources of the sea" but not "the sharing of the sea" itself.[43] This position remained unchallenged in France throughout the years that followed.

Despite the fact that the Community policy structure favoured France in the case of limits, it would be mistaken to suppose that French opinion was satisfied with the CFP in its pre-1977 form. The main interest remained, as it had been in 1970, the protection of the livelihood of French fishermen. In 1975 when economic change started to severely affect the fishing industry and to provoke the unrest that is to be discussed in Chapter 7, it was the EEC market arrangements which were widely condemned as inadequate. The three year reference period of the guide prices, for example, was seen as giving insufficient protection to the producer at a time of increasing inflation. At the same time, the protection against imports afforded by the reference price system was perceived as far too cumbersome.[44] Hence Cavaillé, the Minister responsible, called strongly for the Community to take a more active stance in managing the fish market in the face of external disruption.[45] Though British fishermen were suffering many of the same effects, no major discussion took place in Britain as to the desirability of strengthening the EEC mechanism in this way: the opportunity was not perceived to exist.

4.3 The move towards a revised CFP

After 1977 the debate broadened in that the Community was now responsible for negotiation with third countries and started to become involved in the establishment of conservation measures. The result was that attitudes towards the CFP became more variegated, but the basic themes of British and French thinking remained very much as they had been before. Whereas Britain wanted a lion's share of the fish available, France continued to favour the maintenance of the productive potential of her fishing fleet.

In Britain it was recognised that joint Community action to exclude those states like the Soviet Union which refused to offer reciprocal access had been valuable.[46] However, the question of access to British waters by other Community countries continued to rankle. The Chairman of the WFA expressed a view which remained the dominant one:

> "Every mile that we were prepared to give up below 200 would represent a concession by Britain to all the partners. Here is the true test of the Community spirit."[47]

The result was that British demands changed only very gradually. By June 1977, John Silkin the Minister of Agriculture was calling for a twelve mile exclusive zone with 'dominant preference' in areas up to 50 miles from the coast.[48] Two years later in May 1979 when the Conservatives were returned to power the demand was vaguer but still insisted on an 'adequate' exclusive zone with a 'considerable' area of preferential access beyond.[49]

Moreover, the debate about limits became linked to a new one about Community conservation measures. During 1977 the EEC Commission took over the tasks of NEAFC and began to propose TACs for the species within the Community's 200 mile zone as a way of protecting the resource. It was faced with an initial British argument that because 60% of the Community catch came from British waters, Britain was entitled to 60% of the TACs. This was soon reduced to 45% but there was a major gap between the British claim and the first Commission offer of 31%.[50] Combined with her stance on 'dominant preference', Britain's position on quotas ensured that she was to remain in a minority of one until 1980, when movement towards a settlement started to be made.

Throughout this period British opinion was almost universally obsessed with the inadequacy of the EEC's offers. There was a widespread feeling that TACs were quite useless as a mechanism for controlling catches. Over-fishing was attributed to the predations of the continental states and perceived as only controllable by the legal pressures of the coastal state. In the Trade and Industry Subcommittee, the Fisheries Secretary for Scotland argued that excessive catches were "something for which we can largely blame other countries"[51] and the Director-General of the BFF commented that the behaviour of their fishermen made a mockery of the quota system:

> "...they find it to their advantage to evade the reporting system so that even if the Member State Government is anxious to produce accurate statistics, because evasion is practised, their figures are not even as accurate as they would like them to be."[52]

Hence it was claimed that more than quotas was needed to control fishing. As only the coastal state had the resources to enforce rules, it should be the one to have the role of limiting catches through its own laws. As one report put it, "a pre-requisite for the conservation of fish is that self-denial must be rewarded or made compulsory."[53] TACs alone could not do this and hence the enthusiasm for the unilateral conservation measures which brought Britain into conflict with the European Court of Justice.

Throughout this turmoil over limits and conservation, France maintained her position on the issue of 'historic rights', pointing out that Britain was claiming more than she had ever caught in the EEC zone and more than she had the ships to catch.

> "We, in France, cannot see how the British ... will be able to exploit all the resources they claim to monopolize. Why, moreover, exclude the French by arbitrary decision from these waters, when they have the men, the boats and all the essential know-how? Why not accept the real meaning of the Community?"[54]

However, French enthusiasm for 'equal access' was not without its inconsistencies during this period. The prospect that Spain might join the EEC was to provoke a defensive reaction which bore a distinct resemblance to the attitude of Britain towards the other member states. With French support, the Community negotiated an agreement with Spain which imposed increasingly severe limitations on the level of Spanish fishing activity in the Community zone, in particular, the Bay of Biscay. The hake quota which stood at 14,600 tonnes in 1977 fell to 8,300 tonnes in 1983, while the number of licences available to Spanish vessels declined over the same period from 266 to 123.[55]

Even the Spanish issue, however, reflected the traditional French protectionist response which continued to permeate her attitudes towards the CFP. The conservation policy of the Community proved to be an area where the French were not eager to press for measures beyond the TACs. There was considerable dismay when in the summer of 1979 Walker, Silkin's successor at MAFF, introduced and enacted regulations on the mesh size of nets used for catching Norway lobster. The arrest of a number of Breton trawlers off the South Wales coast provoked considerable unrest in South Finistère and the Minister responsible, Le Theule, wrote to Walker complaining about the unilateral character of the measures taken. It was, though, more than simply a question of disliking unilateral action.[56] There was considerable suspicion in France about the effect on jobs of strict conservation measures, a suspicion reflected in this case by the willingness of the government to contribute to the costs of the vessels arrested. To put the matter another way, there was a general belief that 'biological criteria should not blind policy makers to socio-economic ones.'[57] This was not to say that limits on catches could or should not be introduced but that they should be accompanied by financial help to keep the fishermen in business.[58] As we shall see in Chapter Five, this was to have a marked impact on the kind of aids that were considered legitimate as a form of government intervention.

At the same time, the French continued to press for improvements in the market management arrangements. Following the outburst of protest by French fishermen in August 1980 (to be discussed in Chapter 7), Le Theule wrote to Gundelach, the Danish Commissioner responsible for fisheries, 'a memorandum on the common fisheries policy', in which he called for

> "a reshaping of the common market organisation, the mechanisms of which, agreed in 1970 at a time of relative plenty, have shown themselves to be less and less suited to cope with the situation of scarcity which now faces the Community's fishing industry."[59]

The result of this pressure was that new arrangements were eventually adopted in December 1981. These included the possibility of an accelerated procedure for blocking imports, financial support for the stocking as well as the destruction of surpluses, and a new degressive subsidy for withdrawals, all of which were designed to bring about greater order to the market in fish as the French desired.[60]

Similarly, it was part of the deal which emerged between 1981 and the beginning of 1983 that further help should be made available to assist in the restructuring of the Community's fleet. Again France was following the pattern which had been set in the early 1970s in seeking such support. Le Theule made it clear in 1980 that he wanted the Community to have a stake in his country's efforts to redeploy the fishing fleet and about £10 million were found to support 'exploratory fishing' and 'joint ventures'.[61]

This pattern of a deal which reflected the major concerns of the respective countries also proved accurate in the case of Britain. Despite some interest in the market arrangements, particularly when imports flooded in at the beginning of 1981, Britain's major efforts continued to be directed towards obtaining the best possible terms on access and conservation. Walker, with his Minister of State, Buchanan-Smith, was involved in an enormous number of meetings between 1979 and 1983 with the other states, and not just in Brussels. Over the period they had as many as 47 bilateral meetings outside the formal Community framework in order to persuade the other governments to accord Britain a better deal.[62]

The result of the ministers' efforts was that Britain obtained 37.3% of the stocks available in European waters, as compared with the 31% that she had been offered in 1977.[63] On the question of access, an effective twelve mile limit, with the recognition of certain 'historic rights', was established. Other Community vessels could not fish 'up to the beaches' as the 1970 accord allowed but preferential arrangements beyond the zone of exclusive access were not very substantial. They were restricted to a licensing system to be applied in a box extending about 30 miles out from the Orkney and Shetland islands.[64] As for conservation measures other than the quotas, the desire to strengthen enforcement procedures was reflected in agreement on a British proposal to set up a Community inspectorate.[65] Thus Britain, too, had some reason to be satisfied with the deal, even if Peter Walker's description of it as 'superb' was not universally appreciated.

5. Conclusion

It has been suggested in this chapter that economic change in the fisheries sector was accompanied by three developments: the breakdown of the pre-existing structure for managing the issue; the acceptance of a new structure posing particular problems for state autonomy; and the formulation of different responses to change within that structure. Management of the fisheries sector ceased to be relatively straightforward and uncontroversial and became a much more complex task where the interrelated aspects of the problem were clearly visible and the source of extensive political pressure. At the same time, Britain and France responded in quite distinct ways to the new forum into which the issue was cast. For the former, the issues of access and later, conservation were uppermost, for the latter, the system for managing the market and structuring the industry was the dominant concern. Though the issue acquired increased importance in both countries, it did so for different reasons, determined by the aspect of the new status of the issue which was central to thinking in the country concerned.

However, the existence of these different responses raises the question of why they occurred and what prevented a reorientation. Why did Britain not take a greater interest in market and structural issues and what made France give them such a high priority? It will be argued in the following chapters that the answers lie within the entities 'Britain' and 'France', and can be found by considering the character of the relations that developed between government and fishing industry over the period of this study. Political interdependence and economic change generated the conditions for the choices to be made but they cannot themselves explain them: for this an understanding of the interaction between state and society is necessary. To start to achieve this understanding, the next chapter will consider the organisational framework linking governments and industry in the fishing arena in both countries and the evolution of that framework in the 1970s and early 1980s.

4 The organisational networks

1. Introduction

In the previous chapter it was suggested that both France and Britain were subject to certain constraints in the development of their fishing policies as a result of membership of the EEC but that these constraints did not lead to them pursuing the same policies. The differences between them remained at least as apparent as the similarities in the way they devoted their energies to contrasting priorities.

As a prelude to understanding these differences the present chapter will look at the character of the relations existing between goverment and fishing industry in the two countries. The first section will outline the basic characteristics of the relationship between the state and the fishing industry in the two countries and the different institutional forms that that relationship took. The second section will consider the nature of the criticisms that were directed at the existing framework during the 1970s and the ways in which the basic framework was adapted to take account of those criticisms. The general conclusion will be that contrasting pluralist and corporatist logics operated in Britain and France and that these logics were critical in determining the range of possibilities within which responses to change could take place.

2. The basic framework

The difficulties that the fishing industries in Britain and France encountered during the 1970s were mediated by very different organisational arrangements. In France the state had always exercised a tight form of supervision or 'tutelle', which served both to separate the fishing industry from the rest of society and to act as an important force for homogeneity inside the industry. The system in Britain, by contrast, was marked by a more fragmented type of state involvement, which did not involve an active attempt to set a boundary between the industry and society in general or to limit its heterogeneity. To illustrate the difference this section will look in turn at the general character of the state's role in the two countries and the impact of that role on the particular institutions, which were designed to represent the interests of the industry.

2.1 The role of the state

The origins of the French tradition of supervision of the fishing industry can be traced back to the latter part of the 17th century.[1] At that time Colbert, the Chief Minister to Louis XIV was short of men for the navy. Instead of using the press-gang, as in Britain, he decided to establish a system of conscription, entitled 'inscription maritime'. Under the system, any 'marin', whether he was a fisherman or on a merchant ship, could be enlisted as and when he was needed to man the king's ships. The impact of the establishment of such a link between state and fishermen extended well beyond the navy's requirements for manpower. Fishermen were clearly separated from the rest of society by what Colbert did: they were all registered by the state receiving a number and record card ('fiche matriculaire'), which followed them throughout their working lives. Over time the obligations of service in the navy grew less important but the system became firmly associated in the minds of fishermen with the help that the navy could offer in case of difficulty at sea. So favourable was the industry's view of the system that when in the 1960s it was argued that the character of modern fighting ships made 'inscription maritime' outdated and unnecessary, there was uproar and De Gaulle himself was invited to intervene. He quashed government plans to abolish the time-honoured arrangements and allowed them to continue in an only slightly modified form.[2]

However, there is more to the state's role than that of the registration of 'marins'. Every fisherman has to go through a training course before he can be registered. These courses are run by special technical colleges which are managed centrally by the state-sponsored Association de Gérance des Ecoles d'Apprentisage Maritime (AGEAM). The combination of a particular regime under state supervision is repeated in the nature of the benefits that all fishermen receive once they are registered. As early as 1709 a public body, the Etablissement National des Invalides de la Marine (ENIM) was set up to provide social support for all 'marins'. ENIM has survived to this day and continues to guarantee special treatment for fishermen. Under ENIM rules, for example, it became possible for a fisherman to get a pension at 55, when the age under general social security legislation was 65.[3] At sea, too, fishermen have been aware of the benefits of close ties with the state. Vessels sailing far afield were regularly guaranteed medical and other help from support ships, organised by the industry but backed out of public finance. Ironically, British ships were regularly obliged to turn to these vessels for assistance in that they were rarely accompanied by support ships, except in the special circumstances of the 'Cod Wars' with Iceland.[4]

British arrangements in all these areas have been much more decentralised with much less stress on the separateness of the fisheries arena. Medical provision at sea, for example, has been based on the principle of someone on board having responsibility for the treatment of injuries. Education, too, has seen only limited attempts at central direction: the level and character of provision depends chiefly on local and regional initiative, at the prompting of the semi-public, White Fish Authority (WFA), rather than the central Ministry.

Similarly, the notion that fishermen should be subject to a common system of social benefits distinct from those of the rest of the population is very underdeveloped in Britain. Fishermen in the inshore and deep-sea industries, for example, have never been considered as directly comparable in terms of the character of their employment. The former are categorized as self-employed, while the latter, always employed on a casual basis, have been placed on to a special register of the Department of Employment, from which they are removed if they do not go to sea for six months.[5] It is true that inshore men have certain privileges: they are allowed to stamp their own national insurance cards and pay a special social security rate between that of the self-employed and the employed.[6] However, these remain derogations from a general regime applicable to all rather than the embodiment of a self-contained set of rights and duties for a particular professional category. In this sense the establishment and maintenance of the system of 'inscription maritime' underlines the French state's effort to separate the fishing industry from the rest of society in a way which has found no echo in Britain.

The situation becomes still clearer if we look at the way in which the state's relationship with fishermen has been channelled through governmental institutions. The fragmented character of the treatment of fishermen in Britain is underlined by the way in which responsibility is split between departments. Within England and Wales it is the Ministry of Agriculture, Fisheries and Food (MAFF) that is responsible, in Scotland the Department of Agriculture and Food in Scotland (DAFS) and in Northern Ireland the Department for Agriculture in Northern Ireland (DANI). The fact that each of these institutions is concerned to enforce legislation in the waters adjoining its own areas of responsibility means that fishermen can be faced with different sets of civil servants depending on where they are fishing. The Scottish boats that fished for mackerel off the Cornish coast, for example, were obliged to deal with MAFF as well as their parent department, DAFS. The very fact that two ministries were involved, ministries with different priorities, traditions and reputations, illustrates how hard it is for British fishermen to conceive of the state as anything other than fragmented.

Such a fragmented view is out of the question in France where fisheries have always been under the umbrella of a single ministry, in general, the Ministry of Transport. More importantly, within the Ministry, fisheries has traditionally been part of a separate set of maritime-related sections. Three directorates concerned with the commercial fleet, the administration of all 'marins' and ENIM functioned alongside the Fisheries Directorate in a structure entitled until 1978, the General Secretariat of the Merchant Marine (SGMM).[7] Moreover, within the SGMM a separate cadre of civil servants was developed with their own equivalent of the 'grandes écoles', the Ecole des Administrateurs Maritimes (EAM), based in Bordeaux. The shifting pattern of administrators in Britain is quite alien to France where the products of EAM may move between the maritime sectors but few outsiders come to disturb the privileged link between them and the sectors, including the fishing industry, for which they are responsible.

The importance of the maritime administrators becomes still clearer when one goes from Paris to the regions. As the Economic and Social Council (CES) pointed out in its study of the fishing industry in 1976, the prefects traditionally had no authority over or responsibility for the work of these officials. Just as the fishermen themselves were linked to the navy by the system of 'inscription maritime', so the maritime administrators had a quasi-military status, which put them outside the reach of the prefects.[8] They enjoyed considerable independence both at the regional level, implementing their own division of France into four and then six directorates (cf p 72 above) and at the port level dealing directly with 'marins' through the committee structure to be discussed below.

The comparison between the local maritime administrators and port-based fisheries inspectors in Britain is useful in elucidating very different conceptions of the state's relations with a sector of society. In both cases, there is close contact with fishermen based on shared maritime backgrounds but the nature of that contact is very different. The main concern of the British inspectors is to collect fishing statistics and to administer laws on quotas, net sizes, licences, etc., in other words, to deal with fishermen in their capacity as catchers of fish.[9] By contrast, the French administrators are dealing with fishermen both in a wider context and over a wider range of issues. They are responsible for law enforcement in respect of all maritime matters, including merchant shipping and pleasure craft as well as fishing vessels. At the same time, they are concerned with more than the catching side of the fisherman's life. They issue and manage the 'fiches matriculaires' for enrolling 'marins', they look after their training, they act as arbitrators in disputes, they even have the job of informing wives when their husbands are lost at sea.[10] In other words, they come to represent an integral part of the fisherman's life and to personify the state's 'tutelle' over it. The British fishery inspector has a much less wide-ranging role, the other aspects of the fisherman's life being the concern of separate departments like the Departments of Employment and Social Security or other non-state bodies like technical colleges. The British state within the fisheries sector is therefore confirmed as much more fragmented, and much less concerned to establish a corporate identity for this sector of society, than its French counterpart.

2.2 Representative institutions

The institutional mechanisms which have been developed for representing the interests of the industry in the two states are also quite different. Whereas in France those interests have been channelled through a set of state-sponsored institutions, in Britain diversity of interest has been allowed full expression with minimal state interference.

The representative structure in France took shape in the 1930s at a time of severe economic difficulty for the industry. It turned to the government for help and the result was the creation of a series of port-based fisheries committees. During the Second World War, under the Vichy regime, the arrangements were extended and then, after the Liberation, suitably amended and codified under Decree No.45.1813 of 14 August 1945.[11] The '1945 Decree', as it is still regularly referred to, formed the basis of the representation of the fisheries' interests throughout the post-war period. Under it, the state established a whole series of committees covering the local, regional and national arenas each designed both to consult the industry and to help to manage the fisheries sector. However, the industry was not left free to determine how these roles should be carried out. The decree laid down as basic principles that membership be compulsory, 'interprofessionel' i.e covering all sectors of the industry and 'syndicale' i.e involving the unions active in the industry. At the same time, the state was guaranteed access to the deliberations of this consultative structure.

In Britain, by contrast, similar economic pressures produced a very different state response: a clear distinction between the representative and management functions was maintained. Two statutory, semi-public bodies, the Herring Industry Board (HIB) and the White Fish Authority (WFA), set up in 1935 and 1951 respectively, were established to deal with the economic difficulties of the industry, while the mechanisms of consultation within the industry were left untouched. No attempt was made to integrate the range of interests involved, nor to give the state a direct role in the deliberations of those interests. The contrast becomes still clearer by looking more closely at what the structures look like in the two countries.

In France the peak of the consultative structure is the Central Committee for Sea Fisheries (CCPM).[12] Within this committee there are currently 75 voting members made up as follows: 40 from the catching sector, 31 from the processing and distribution sectors and 4 from the shellfish industry. It is specifically laid down that all parts of the catching sector be represented: inshore and deep-sea industries, as well as owners, officers and crews.[13] In addition, 11 civil servants have the right to attend, though not the right to vote: they come from the ministry directly concerned with fisheries as well as those with an interest in the sector, including the Ministries of Finance, Supply and Industrial Production.[14] The President of the Committee may himself be a civil servant, like Dubreuil, who took office in the autumn of 1974 and was still in post at the end of the period of this study. If not from the industry, the President too has no vote.

This lack of voting rights does not mean a lack of state influence. The civil servants on the committee do have the right of veto, while the Minister responsible for the industry can play an important role. He is responsible for the annual appointment or reappointment of the President and he must also formally confirm the decisions that the whole committee takes. If he takes exception to its actions, he has the power to modify its reports or even to suspend its operations.[15]

The powers of the CCPM are more those of consultation and coordination than of direct management. It has the task of watching over the activities of the local and regional committees, of proposing ways to improve the standard quality of fishing vessels, methods of fishing and marketing and more generally, of investigating any measures that could benefit the industry as a whole. However, it does have the right to propose the level of an 'ad valorem' duty payable by all those selling fish to cover the costs of administration of the various committees, as well as the maintenance of a fund for promoting the consumption of fish.[16]

This last point offers a direct comparison with the WFA and HIB.[17] All members of the UK industry, catchers, merchants and processors, were obliged from the outset to contribute to the running of these bodies and to the encouragement of fish consumption by means of a levy on fish, equivalent to that payable in France. However, there was no membership link established between the WFA and HIB, on the one hand, and the industry on the other. Indeed industry representatives were specifically excluded from belonging to their governing boards on the grounds that this would undermine the independence of such semi-public bodies. By the same token, no state representatives could sit on the boards, thus isolating their members from the two forces (i.e state and industry) around which the French structure revolves.

However, the two British bodies were given major financial powers 'to reorganize, develop and regulate the industry' which went beyond anything available to the CCPM. They were responsible for the award of grants and loans for the construction or improvement of vessels, transmitting money coming from the Treasury at a slightly higher rate than that requested by the Treasury for repayments.[18] To do this, they established a set of regional officers whose job it was to adjudicate on the requests made by individual catchers. Such requests in France remained outside the purview of the CCPM and firmly under ministerial control, every dossier having to be submitted to Paris for a decision.[19]

However, the HIB and WFA were given no role in the day-to-day management of the fishery: like the CCPM, they could study ways of improving fishing but they could not establish, even at regional level, rules by which fishermen were legally bound in the pursuit of their livelihood. By contrast, beneath the CCPM in France, the 1945 Decree set up bodies which reflected the national committee in composition, but were given formal powers to establish legally-binding rules.

Thus at the national level, a set of nine Interprofessional Committees (comités interprofessionels) was established to cover each of the sea products gathered by French fishermen.[20] These committes are structured similarly to the CCPM: there are representatives from all the different sections of the industry; there is a Ministry representative who has a right of veto, though not the right to vote; and there is also a President, appointed by the management board rather than the Minister, as in the case of the CCPM.

Despite being organised nationally, many of these committees have a strong regional orientation. In the case of herring, for example, it is only ports on the North coast that are involved and even when more than one region is concerned, as in the case of sardines or demersal fish, there are subcommittees for each of those regions. As a result, they are more directly concerned with the specific problems of sections of the industry than the CCPM and this closer link is confirmed in the powers that they possess.[21] They were empowered by the 1945 Decree to fix the opening and closing dates of fishing seasons, to determine the number of vessels allowed to fish, to regulate the number of voyages, to establish minimum quality standards and to set up any other bodies that might help vessels to operate more effectively.

The same kind of stress on direct management, though not through legally-binding rules, can be seen in the provisions relating to Local Committees (comités locaux) or Regional Committees (comités regionaux), where there is a grouping of Local Committees.[22] They were given the task of creating and managing collective services, such as cooperatives and auctions as well as organising the share-out of fuel and supplies for vessels. In the tradition of 'inscription maritime', they were also invited to take initiatives in the training of 'marins', and more generally, to seek improvements in social conditions. The state's commitment to such a role extending beyond consultation is confirmed even at this local level by the right of the maritime administrator in the area to attend and to exercise a veto, should he choose.

In Britain, by contrast, the presence of the state at regional or local level is confined to the fishery inspectors, who as indicated above (p122) have a less wide-ranging set of links with the industry than their counterparts in France. At the same time, their task is to implement regulations and not to devise new ones. The only British bodies that could be remotely compared to the Interprofessional, Regional and Local Committees are the Sea Fisheries Committees which exist in England and Wales, though not in Scotland. They are committees of one or more County Councils, who have the power to make bye-laws, subject to ministerial approval, prohibiting fishing in coastal waters and imposing limits on the size of fish caught. However, they have never formed as integral a link between state and industry as the 1945 Decree institutions have done in France.[23]

To understand more fully the importance of the 1945 institutions in France and the extent of the contrast with Britain, it is necessary to return to the principles of membership mentioned earlier, namely that the French structure is 'interprofessionel' and 'syndicale'. It is part of the philosophy of the 1945 institutions that all sections of the fishing industry should be obliged to sit around the same table. This unitary conception of the industry is absent in Britain, not because the characteristics of the industry are different but because there has been no concerted attempt on the part of the state to overcome the problem of differential access of groups within society to the state apparatus. Within France the recognition of this kind of problem was what led to the establishment of the CES in 1925, following the demand of the General Confederation of Labour (CGT) to have direct links with government.[24] Thus the CCPM structure formed part of a wider tradition, a tradition totally absent in Britain.

Apart from obliging the catching sector to sit down with merchants and processors, the 1945 Decree was directed to ensuring access for both the inshore industry and the unions. In both countries the deep-sea sector with its company-owned boats has always enjoyed natural advantages of scale and concentration. Thus in France, even after 1945, the owners of 'la pêche industrielle' maintained their own separate organization, based in Paris, the Union of French Fishing Boat Owners (UAP), which was never matched by an equivalent body for 'la pêche artisanale'. For the inshore boats, the 1945 institutions guaranteed formal parity, with their 20 seats matching the 20 of the deep-sea industry even if, as some argue, the deep-sea industry maintained its effective dominance, without appearing to.[25]

In Britain, by contrast, until the 1970s, the dominance of the deep-sea industry was in no way concealed. It grouped together the owners not on a national basis but in bodies which reflected the governmental split between MAFF and DAFS. In England and Wales, it was the British Trawlers' Federation (BTF), based in Hull, which represented the owner interest and in Scotland,it was the Aberdeen-based Scottish Trawlers' Federation (STF). As for the inshore industry, it remained completely fragmented, rarely able to transcend the differences between individual ports or regions. In Scotland, there had been a national grouping in the post-war period but it broke down in 1948 in the face of divergent interests.[26] In England and Wales, the only wider form of association was the Fisheries Organisation Society Ltd., a body almost entirely financed by the government.[27] However, it was not strictly a representative of the inshore interest but an institution set up in 1914 by the government 'to foster the propagation of cooperative principles amongst inshore fishermen'. In other words, no specific attempt was made by the state in Britain to overcome the low level of access of the inshore industry to the policy process as happened in 1945 in France: the whole framework of consultation was much less formal, much more haphazard. As one Fisheries Secretary at DAFS put it: "we tended to follow the line of consulting everybody who we thought had something to say."[28]

A similarly detached attitude can be observed in British attitudes towards union involvement in representation of the industry. As there was no prescribed right to consultation but only consultation by invitation, it was possible for representatives of the Transport and General Workers' Union (TGWU), the only union with members in the catching sector, to feel excluded, even if not actually to be so.[29] It is true that membership was never high and did not extend outside the deep-sea industry. Even amongst deep-sea employees, there was a clear division between the crews, on the one hand, and the officers and captains, on the other. The latter were generally represented by port-based Trawler Officer Guilds (TOGs). As for the inshore industry, it remained firmly opposed to any union encroachment: in Scotland, as we have seen (see p 75 above), the fishermen went so far as to move their operations from Aberdeen to Peterhead to avoid having to pay for the unloading of their ships by union labour.

At first sight, the unions in France would appear to suffer

from at least as many handicaps as the TGWU. Their membership has also never been high and been very much concentrated in particular ports and regions. What is more, there is not one but three unions competing for membership: the CGT, the French Democratic Confederation of Labour (CFDT) and the French Federation of Professional Maritime Unions (FFSPM). In theory, the wider conflict between the two leftist unions in France, the Communist-inclined CGT and the Socialist-inclined CFDT is reproduced in the fisheries sector and complicated by the involvement of an independent union, the FFSPM. However, the 1945 Decree wished to rectify the arrangements established under the Vichy regime, where only one union was permitted to represent the employees in the industry. This overtly corporatist practice was modified by introducing union competition but in effect, the three unions mentioned were given the monopoly right to nominate the 20 employee representatives of the industry (10 for the deep-sea, 10 for the inshore) on the basis of their relative strengths in work-place elections. In this way, whatever their membership levels, the unions were firmly incorporated into the consultative structure and given a direct stake in its maintenance. Such were the consequences of the French state shaping the representative institutions in the fisheries sector, a process without parallel in the British context.

3. The effects of a changing environment

In both countries, economic change in the 1970s provoked dissatisfaction with the basic institutions available for representing the interests of the industry. The criticisms had two main elements, one questioning the effectiveness of the existing institutions to cope with new challenges, the other drawing attention to the overall institutional strength of the fisheries sector in relation to other sectors of the economy. The different traditions which had prevailed hitherto meant, however, that the content of the criticisms was very different in the two countries as indeed was the response that they evoked. In Britain the moves forward were firmly in the pluralist tradition with its stress on the development of competing, voluntary organisations, restricted to the transmission of grievances; in France, the corporatist ideology of a unified sector was extended by the creation of new bodies expressing that unity, with an emphasis on the economic protection of the industry. Here the two countries will be considered in turn.

3.1 The British experience

As the 1970s progressed, there was a growing recognition of the disadvantages of the fragmentation of the fishing industry in Britain. From a governmental point of view, the process of consultation became increasingly difficult and compared very unfavourably with other sectors in its complexity. Whereas, the food and drink industry, for example, has 18 branches but represents its views through a single body, in fishing practically the reverse is true, with one branch, the catching sector, sprouting a plethora of representative institutions.[30] Hence civil servants openly favoured closer links between the various institutions. Kelsey, the Fisheries Secretary at MAFF from 1976-1980, commented to the Trade and Industry Sub-Committee: "there is scope for shortening lines of communication; concerting a line of action would certainly make them more quickly effective in these days of fast-moving decisions."[31] And when the Committee reported, it went as far as the voluntarist tradition would permit, in backing Kelsey's line: "we think that the industry should be encouraged to consolidate its representative institutions to the greatest extent that it finds practicable."[32]

The industry itself did respond to this kind of urging: it recognised the need for increased political strength through greater unity. As early as 1973, six local associations in Scotland combined to form the Scottish Fishermen's Federation (SFF) and it expanded during the 1970s to include nearly all the inshore boats in Scotland. With earnings of about £50,000 per year, it was able to set up an office in Edinburgh with a small full-time staff and to start producing a newsletter, informing members of its lobbying activities. Despite strong disputes between the constituent members, the SFF remained a permanent feature throughout the period of this study.

Even within the deep-sea industry the threat of expulsion from Icelandic waters led to the merger of the BTF and the STF in September 1976. The British Fishermen's Federation (BFF) incorporated the high level of organisation and professionalism of its predecessors, as epitomised by its Director-General Austen Laing. However, it was faced with the prospect of decline, as the deep-sea fleet itself shrank. In May 1979 the newspaper of the Federation, Trawling Times was wound up and in 1983 Laing himself retired. Finally, in March 1984, the Federation was disbanded, when there were hardly any interests left to represent.

Within the inshore industry in England and Wales, the difficulties encountered at the beginning of 1975 led to pressure for a single voice. The search was complicated by the existence of the FOS, which claimed already to be doing this job. However, quite apart from its financial dependence upon government, its members were concentrated regionally in the South West, even though its headquarters were in Surrey. In the North, in particular there was no confidence in the FOS and in May 1977, the National Federation of Fishermen's Organisations (NFFO) was set up, covering England north of a line from Grimsby to Fleetwood. It slowly increased its influence at the expense of the FOS and by 1979 it was actively considering joining the SFF and BFF in the European pressure group 'Europêche' at a cost of £1,333 per year, a clear indication of the relative health of its finances.[33] By 1983 its full-time officials had become regular attenders at the Council meetings in Brussels, even though it remained overshadowed by the larger Scottish body, the SFF. By contrast, the FOS failed to survive the cutting of the government grant in 1980 and went out of existence in May 1982.[34]

However, it was not simply a question of how best to unite to transmit the views of the industry to government. There was also considerable debate as to the quality of the governmental institutions and their capacity to defend the industry adequately.

This debate revolved around the ministries responsible for fisheries and the role of the WFA and HIB.

Criticism of the governmental apparatus concentrated, in particular, on MAFF in London: DAFS in Edinburgh received a better press, being seen as closer to the fishermen geographically and in spirit. The two fishery divisions of MAFF, by contrast, drew fire from every quarter for their level of competence, the nature of their allegiances and their ability to defend the industry. There was a general feeling that these divisions were weak both in number and quality, summed up in the Commons by John Nott, the Conservative member for St.Ives:

> "Having had ten years' experience of trying to help the fishing industry in my constituency, I know of no more feeble department in Whitehall than the fisheries side of the Ministry of Agriculture, Fisheries and Food. However hard one tries, one experiences difficulty in getting positive and constructive action out of that department."[35]

This jaundiced view of MAFF was compounded by a feeling that the department was biassed in favour of the owners: its officials were described to the author as 'the lackeys of the English deep-sea companies',[36] a charge which underlines the geographical as well as the functional fissures in the British industry.

Perhaps most persistent was criticism of the ability of the MAFF fishery divisions to defend the industry. In part this was linked to the traditional British complaint of the lack of continuity in the civil service, where fish can be just an interlude between the Treasury and tourism.[37] More important was the question of the relative weight of the fisheries interest within MAFF. Johnson, the Hull West MP, argued in the Commons in 1975 that "fishing is the Cinderella in comparison with its massive partner agriculture in that office in Whitehall."[38] For him this was confirmed by the fact that at the Dublin Summit of EEC Heads of Government of March 1975, which agreed on the renegotiated terms of British membership of the Community, fishing was not even on the agenda.

The result of this final criticism, in particular, was to create some pressure for a separate Ministry of Marine Resources, within which it was thought that the fishing interest would not be swamped. However, it never became a major issue. Silkin, the Minister responsible for MAFF, dismissed it as irrelevant, maintaining that the important thing was the "very large expertise", built up over many years by the Ministry.[39] Thereafter the issue effectively disappeared, though this is not to say that there were no institutional changes within the government machine. By 1979 the number of fisheries divisions within MAFF had expanded from 2 to 4, reflecting the concerns of the government as stressed in the previous chapter. Thus new divisions were set up to deal with the Law of the Sea and fishery limits as well as fishery protection and quota management.[40] The fishing industry, however, remained firmly anchored within MAFF.

The final area of criticism and change was the activities of the WFA and HIB. These bodies had always been the subject of some dispute within the industry. Fish merchants, in particular, were dissatisified with having to contribute to the running of bodies whose spending was concentrated on the activities of the catching sector.[41] Indeed even within government there was uncertainty as to whether they were the appropriate institutions for the job. As long ago as 1961 the Fleck report had recommended that they should at least be merged and in the early 1970s, Prior, the Minister of Agriculture, had wanted to go further and to dismantle them altogether.[42]

These difficulties assumed larger proportions as the problems of the industry increased. The two bodies were caught between their clients, the industry and their creators, the government. They found it hard to develop a new role in the context of economic change, particularly as the provision by them of grants and loans for boat building and improvement was undermined by a drastic reduction in government money. In his evidence to the Trade and Industry Sub-Committee, Meek, the Chairman of the WFA, suggested that by the late 1970s, the most important functions of the Authority were research and development, the development of fish farming and publicity, arguing that these were jobs that "the industry cannot do very well for itself."[43] He implicitly acknowledged the limited nature of these functions by favouring an increased role for the Authority, including larger funds for financing advertising and the devolution of licensing powers to it from the government.

The problem with developing the powers of the WFA and HIB was the fishermen themselves. They resented any notion of excessive independence on the part of the two bodies. When the Chairman of the HIB suggested in 1975 that the fishermen should not press for an exclusive 50 mile limit until the resolution of the EEZ issue at UNCLOS III, he was roundly condemned. As one fisherman put it:

> "We feel that a chairman of the HIB should be a liaison officer between the Government and the fishermen and as such, his opinion should represent the trade as a whole. He should never be placed in the position where he is obliged only to speak of the Government's policy."[44]

Perceived as 'liaison officers', the WFA and HIB could hardly expect the fishermen to favour an increase in their powers. The incentive for such an increase was in any case lacking. More publicity, for example, could only come if the levy on landed fish was higher but that levy could only be higher if the fishermen themselves increased their contribution, something they were not enthusiastic to do.

Furthermore, the fishermen themselves were given the opportunity under EEC legislation to extend their own powers by establishing producer organizations (POs) (cf p 102 above): these effectively weakened the WFA and HIB by allowing producers to take unto themselves day-to-day marketing and management.[45] Very quickly some POs were able to develop an organisational sophistication that made the ordinary representative institutions look very primitive. The £50,000 per year earnings of the SFF with its full time complement of two (one man and his secretary) were rather insignificant when set against the plush opulence of the biggest PO in Britain, the Scottish Fishermen's Organisation Ltd (SFO). In 1979 it had earnings of £300,000 with 12 full-time employees in its Edinburgh office and 18 port officers. Such earnings were similar to the total publicity budget of the WFA for the whole country.[46] Fishermen were still contributing to the WFA by paying the levy on landings but the role of the SFO in supporting prices offered a more tangible benefit than the rewards of advertising, offered by the WFA.

Eventually the merger of the WFA and HIB, mooted 20 years earlier, took place when on 1 October 1981, the Sea Fish Industry Authority (SFIA) came into existence. It was not, however, a body that was given major new powers over the industry in general, or the POs in particular. Rather it introduced a new kind of mood, epitomised by the replacement of the ex-colonial servant, Charles Meek, with the former sales director of Raleigh bikes, Peter Seales. The Board structure was modified with industry interests allowed to join and the ethic of business enterprise was developed with an advertising campaign transported around the country by train. Not all were happy with this break with the past: indeed after mass sackings of staff, Seales himself resigned in 1983 amidst a general feeling that the removal of the old institutions had been far from an unqualified success. Britain seemed better able to find fault with existing institutions than to imbue new ones with a clear sense of purpose: this was not the case in France.

3.2 The French experience

In contrast to British complaints about the fragmented representation of the fishing industry, the debate in France revolved around the ability of the 1945 institutions and the ministry that exercised supervision over them to respond to economic change. The problem was not one of how best to transmit the views of the industry but of how to modify the structure to make it more dynamic. The institutional separation of the industry from the rest of society no longer appeared as a guarantee of privilege but of isolation and neglect.

The thrust of the main criticisms was contained in the report prepared for the Economic and Social Council (CES) by J. Martray, following the unrest in the industry at the beginning of 1975. The CES underlined the isolation of the fishing industry by noting that it was producing a report on that industry, despite the absence of any fishing representative on the Council.[47] It went on to argue that the 1945 institutions were not just isolated from the mainstream of political life but were quite incapable of coping with the new pressures with which they were faced. The CCPM had, according to Martray, been "paralyzed" by the 1975 disturbances, serving only as "a platform for the often conflicting demands of the various groups represented on it."[48] At the same time, the report criticised the SGMM as a relatively ineffective defender of the fishing interest, especially when compared with its equivalent in other countries. It rejected the idea of integrating fisheries into the agriculture ministry, arguing that the problems were different and that the larger interest was likely to swamp the smaller one. However, it did favour the establishment of a Ministry of the Sea, thus putting the idea on to the political agenda for the years that followed.[49]

The response of the government in institutional terms was an implicit acknowledgement of the validity of the CES criticisms concerning the capacities of the 1945 institutions. As early as February 1975 the industry was pressing for a national organisation to control the fish market, comparable to that existing for agriculture, and at the beginning of the following year, the Prime Minister, Chirac, announced the creation of the Market Intervention and Organisation Fund for Sea Fisheries and Shellfish Culture (FIOM).[50] FIOM was given two main tasks, the first of which was to intervene and to help to manage the fish market.[51] This meant support for prices, control of surpluses to avoid their destruction, encouragement of deals between producers and processors, help in the search for export markets and backing for fish publicity. Its second task was to direct fishing towards new zones and species by backing experimental voyages. The exercise of these two roles will be considered in more detail in

Chapters 5 and 8 but here it is necessary to consider the character of the relationship existing between FIOM and the other institutions in government and the industry.

First of all, FIOM is not a semi-public body like the SFIA but firmly under the 'tutelle' of the ministry. Thus in a manner reminiscent of the CCPM the 26 members of its Management Council include 6 civil servants as well as 18 representatives of the industry. Its establishment was partly seen as gaining the benefits of such a 'tutelle', whilst avoiding the bureaucratic rigidities of the Ministry. Thus from the outset it received a large state subsidy (about 57 million FF in 1976), whilst being able to spend that money through relatively uncomplicated budgetary procedures.[52]

Secondly, again unlike the SFIA or its predecessors, FIOM came to establish close links with the producer organisations provided for under EEC legislation. In France the system of POs had already assumed an important level of coherence before the establishment of FIOM. In April 1975, 15 inshore and 3 deep-sea POs, covering 90% of the industry had decided to get together to form a National Association of Producer Organisations, (ANOP).[53] In October 1976, the French state gave exclusive recognition to ANOP and thereafter the ties between the POs and FIOM developed strongly. Both sides shared a common interest in market conditions and the existence of FIOM created the opportunity for further strengthening the role of the POs. As the Director of the Fisheries Directorate in the Ministry put it:

> "The action of FIOM embodies an impetus, a support, a will, at the local level, to establish a system of guarantees in the face of the uncertainties of fish catches."[54]

However, both FIOM, and the POs, did face a problem, namely, their relations with the 1945 institutions. Although the President of the CCPM also chairs the Management Council of FIOM, sources of conflict were clear from the outset. Part of FIOM's income comes from a share of the levy raised by the CCPM and as a result the CCPM effectively supported the encroachment by FIOM upon its prerogatives. The advertising activities of the National Committee for the Support of Consumption of Sea Products (PROMER), for example, came under FIOM's aegis, although PROMER had traditionally been financed and supervised by the CCPM.

Within the CCPM there was a recognition that its legal framework and mode of operation made it more suitable to act as a forum for discussion, where "the major choices ripen", than as an executive body,[55] but there was considerable suspicion as to what kind of role FIOM might play. Would it simply guarantee the losses of the deep-sea companies when they were in trouble leaving them free to reap profits when times were good? Or would it seek to guarantee an increasing level of income for only one category, notably, the 'patrons' of the inshore industry?[56]

Such criticisms were especially strong amongst the union representatives, who had a legally-guaranteed position in the 1945 structure but lacked an equivalent strength in FIOM, despite a presence on the Management Council. They observed, in particular, that membership of the POs, in contrast to the local committees, is restricted to producers, i.e owners:crews, whether in the deep-sea or the inshore industries, are not represented.[57] As a result, they were eager to bolster the 1945 structure: a CGT initiative, for example, led to the setting up within that structure of a National Committee on the Fishing Fleet, designed to voice union grievances over the decline in the number of ships.[58]

However, there was little scope for the 1945 institutions actively to resist FIOM and the POs. Even the Interprofessional Committees did not have the kind of direct powers of market management possessed by the POs. Hence from June 1981 the new French government made increased efforts to integrate the two structures. There was no question of abolishing the CCPM and its attendant committees: Le Pensec, the Socialist Minister, could not contemplate the uproar that would have followed. Hence a more pragmatic solution was devised whereby the head of ANOP became a member of the CCPM and representatives of the individual POs joined the local committees. In this way, the old structure was maintained and the unions received the satisfaction of gaining a say in the running of the POs.[59] The end result is summarized in Figure 6, which underlines the link between the two structures.

As in Britain, however, there was also concern about the weight given to fishing within the counsels of the government. At the time of the CES report with its call for a new Sea Ministry, fisheries was simply one directorate within the SGMM, which in turn depended upon the Ministry of Transport, whose minister was

FIGURE 6 - Organisation plan of the French fishing industry

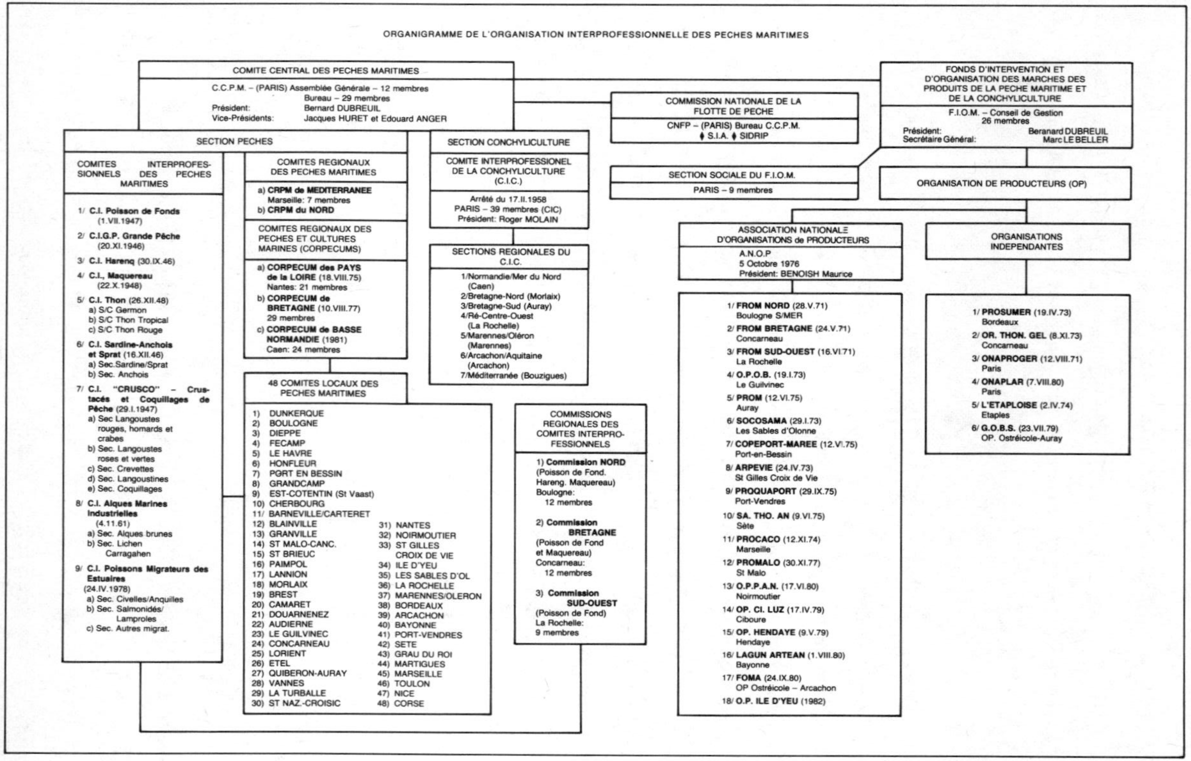

Source: CCPM March 1983

then not of Cabinet rank. Such junior status had already provoked comment before the Martray report[60] and it continued to do so afterwards.

Barre, Prime Minister from September 1976, responded to the criticism by establishing an Interministerial Mission for the Sea and by upgrading the SGMM into a Directorate-General (DGMM)[61] but these moves did not serve to placate the industry. The upgrading of the SGMM to the status of the other Directorates-General of the Ministry of Transport was seen as undermining the specificity of the industry, especially as it also involved the suppression of the separate Directorate concerned with the administration of all 'marins', the 'Direction des Gens de Mer'.[62]

However, the issue did not disappear but re-emerged in still sharper terms during the 1980 dispute, to be discussed in Chapter Seven, and afterwards in the run-up to the Presidential and legislative elections of May and June 1981. The result of the Socialist victory in these elections was the creation of a new Ministry of the Sea under the Breton, Le Pensec. As the organisation plan of the ministry indicates (see Figure 7) it was not particularly radical in conception. The 'Direction des Gens de Mer' was reinstated and the Interministerial Mission was integrated into it but there was no question of defining its responsibilities more widely. Pollution and oil prospecting, for example, were not included within its competences. Nevertheless,

FIGURE 7 - Organisation plan of the Ministry of the Sea

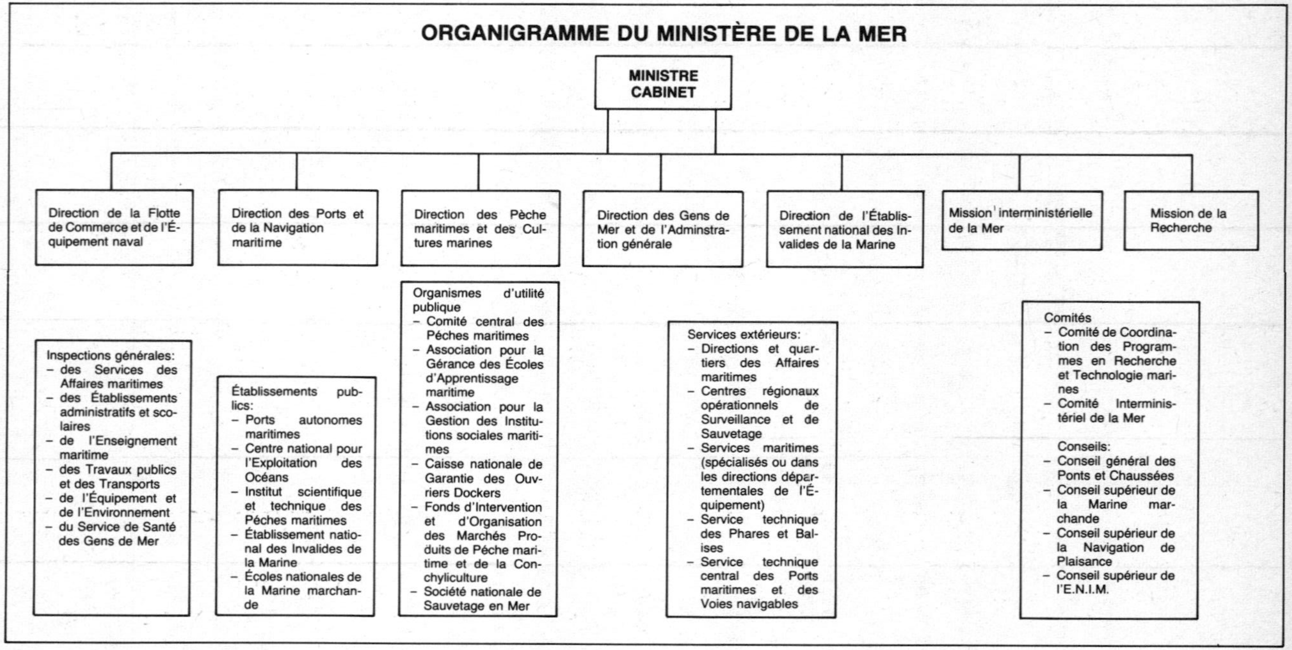

Source: LA DOCUMENTATION FRANÇAISE (1982) Supplément aux Cahiers Français No.208, October-December 1982, Gérer La Mer

it had a powerul psychological effect within the fishing industry. The CCPM commented:

> "Everyone sees in it the long-nurtured hope of at last being listened to, understood and noticed by attentive, open and accessible authorities. This innovation is the promise of all the improvements awaited since 1975, the promise of fishing that is profitable once again, the promise that this essential economic activity and its human environment will not be sacrificed to statistical, technocratic and economic considerations, without any link to reality."[63]

Le Pensec proved a very popular minister in his efforts to fulfil these hopes. He made it his business to attend the annual general assemblies of the CCPM and of ANOP held in October 1981, and was still receiving glowing reviews after a year in office.[64] Indeed in March 1983 when the Ministry was downgraded and put back under the supervision of the Ministry of Transport, Le Pensec's resignation was greeted with considerable disappointment and a feeling that the government had taken something of a step backwards in its defence of the industry.[65] Nevertheless, the various directorates, including that for fisheries, remained intact under a Secretary of State responsible for the Sea (Lengagne, the mayor of Boulogne). Over the period from 1978 to 1983 the precise institutional context within which the fisheries issue was handled had changed considerably (more so than in

Britain) but it still reflected the basic French concern to differentiate 'marins' from the rest of society, a concern dating back to Colbert.

4. Conclusion

What this chapter has demonstrated is that economic change generated demands for institutional change but that the demands were different in character and effect in the two countries. In Britain it was the fragmentation of the representative institutions that attracted attention. The industry itself made some efforts to remedy the situation but it received little more than encouragement from the state, which did not seek to exercise a tighter form of institutional supervision over the fisheries sector. Hence a broadly pluralist system prevailed with continuing competition between freely constituted, voluntary bodies.

In France such competition was dampened within a more corporatist set of arrangements, where membership was compulsory and the institutions state-sponsored. Though it was the ineffectiveness of these arrangements which provoked criticism, the state remained central in determining the shape of change and continued to channel the economic and political energies of fishermen. Hence it is not surprising that the expression of support for the industry, provided by a separate ministry, emerged in France rather than Britain, despite similar feelings of isolation on the part of the industry in the two countries. This pluralist/corporatist dichotomy has pointed ahead to further contrasts between the two countries to be developed in the four chapters which follow. It has emerged, first of all, that the institutions reflected different economic priorities. The creation and development of FIOM illustrates how much importance the French state placed on the control of the internal market as a way of protecting the industry. None of the initiatives in Britain indicated a desire to move beyond establishing a framework within which the industry could pursue its economic activities. The nature and level of the state's intervention in the economic workings of the industry will be pursued in the next chapter.

Secondly, this chapter has suggested that there were different amounts of room for manoeuvre available to the fishing interest. In Britain the freedom allowed by the state to create institutions or not, as one chose, carried with it the possibility of a wide set of strategies of influence. In France the state's 'tutelle' limited the industry's scope for manoeuvre, by channelling its energies through officially-approved bodies. This issue of how the demands of the interest were mediated by its representatives will be taken up in Chapter Six.

Thirdly, this chapter has pointed towards a variable potential for opposition betwen the state and this sector of society. The tightness of the link between the two in the French context left little margin for error, should the structure fail to contain the demands made by the industry. By contrast, the 'distance' of the British state from the industry offered more opportunities to prevent relations with the industry from deteriorating and getting out of hand. The character of direct action taken by fishermen in opposing the state will be discussed in Chapter Seven.

Finally, this chapter has indicated one aspect of the industry's efforts to organise its own affairs, the PO structure. In Britain that structure developed erratically but in France it became almost a mirror image of the 1945 institutions thanks to FIOM's encouragement. A more detailed look at the different facets of self-help will be taken in Chapter Eight.

5 The interventionist perspective

1. Introduction

The present chapter will consider the first of the perspectives outlined in the initial chapter and examine the nature and extent of the state's intervention in the fishing industry in Britain and France. The material will be used to see whether the nature of the changes in the issue area and the general character of relations between state and society in the two countries influenced intervention in the way we might expect. It will be recalled that we started with two assumptions, first that "the industry would demand increased intervention from government to protect it from the effects of the changing environment and that the government would feel itself obliged to cater for these demands" and second "that the 'dirigiste' French tradition and the British aversion to intervention would mean that government efforts to determine the shape of the industry were more developed in France than in Britain" (cf p 30 above). Both assumptions will be found to be confirmed by the evidence presented in this chapter.

The chapter is divided into three sections: the first will suggest that increased levels of support to the fishing industry cannot be compared in purely quantitative terms but need to be understood in terms of the different philosophies of assistance prevailing in the two countries; the second will look at the kind of subsidies that were considered appropriate to deal with the increased costs of the catching sector and the nature of the debate that those subsidies provoked, and the third will examine the response of the two states to their progressive exclusion from the waters of third countries. The main thrust of the argument will be that the behaviour of the state institutions and of the industry in France confirm the view that there, unlike in Britain, "the public interest is not reducible to some compromise between private interests".[1] The political and economic strands of the French state tradition combined in the case of the fishing industry to generate a policy impetus, unmatched in the British context.

2. The basic character of intervention in the industry

There is no doubt that during the 1970s and early 80s, the level of economic support given by Western governments to their fishing industries increased markedly. OECD commented drily in 1981 in its annual review of the fishing sector:

> "In several OECD countries, it is normal for the fishing industry to receive financial aids from the state. However, it should be noted that these aids have become more numerous in recent years."[2]

Both France and Britain were countries that had 'normally' given such aids to their industries. In Britain cyclical depression in the market had led to the introduction of operating subsidies at various times during the 1950s and 1960s[3] and throughout that period the WFA had maintained a generous policy on grants and loans for the building of new ships. Indeed in 1970 one French observer saw the £4 million annual budget of the WFA as guaranteeing a rosy future for the British industry for the rest of the decade.[4] In France, too, there had been significant help with the construction of new boats, whose value to the industry was only voiced openly when they were being phased out in 1973.[5] Furthermore, the social benefits available to fishermen under the ENIM regime (cf p 118 above) had always provided a significant hidden subsidy which knew no parallel in Britain.

However, the oil crisis of 1973 and its consequences provoked a major increase in demands for state assistance. Meek, the Chairman of the WFA, even used the fact that the industry had always been subsidised as an argument for extra support now in the face of its "sharp reversal in fortune."[6] French reaction was equally strongly in favour of state aids, with particular stress being laid on the relative position of fisheries in the economy. A figure of 30 million FF in the budget for fisheries was contrasted with the 800 milllion FF available to help the shipyards.[7] Whether such comparisons were justified is less important than the fact that they were made and that governments responded with fresh subsidies.

There was a marked increase in government expenditure in support for the fishing industry in 1975 and 1976. In Britain, the figure rose from £20.5 million in 1974-75 to £37.1 million in 1975-76, while in France there was a more dramatic rise from 84MF in 1975 to 223 MF in 1976, the latter figure including 57MF for the setting up of FIOM.[8] Thereafter the figures dropped again, though not to the pre-1976 level, until 1980 when the governments of both countries were faced with renewed demands for assistance from the industry. In 1980, the French government accorded aids amounting to 202 MF, while in Britain successive supplementary subsidies of £3 million and £14 million were annnounced in March and August.[9] Moreover, this trend of increased assistance continued in the two subsequent years preceding the EEC settlement. In the Spring of 1981, the British government granted a further £25 million, followed up in October 1982 by another package of £15 million.[10] In France the advent of the new team at the Ministry of the Sea, under Le Pensec, led to a still more marked increase in support. The 1981 figure totalling 389 MF was almost double that of 1980 and the 1982 figure of 350 MF amounted to a sevenfold increase in constant prices over the level of aids accorded in 1970.[11]

The political importance of these figures lay in their relative rather than their absolute size. For one thing, it is notoriously difficult to calculate what is to count as an aid to the industry. Thus in the French case, do the figures quoted above include the deficit caused to the state by the financing of the social benefits of ENIM, a deficit said to match that incurred in direct aids to the fishing fleet?[12] The source used offers no guidance on the matter. However, in political terms, the reliability of any statistics was much less important than their availability. The institutional interdependence, outlined in Chapter Three, fostered the feeling of fishermen in each EEC country that their counterparts abroad were being guaranteed a higher level of support than they were. Thus when the Commission answered questions of members of the European Parliament (MEPs) on the relative costs of fuel for fishermen in the member states, they were inevitably providing ammunition for the debate at national level.[13] During the 1980 dispute, Le Theule, the Minister of Transport, could use the argument that fuel for French fishermen was the cheapest in Europe to try to quell the protests.[14] At the same time, Buchanan-Smith, Minister of State at MAFF, could take a total figure for French and British aids (£23m and £18m respectively) to support the view that the industry's demand for an extra £35m was "extremely large" and by implication, out of all proportion.[15]

What the political debate over relative levels of subsidy obscured, however, was the basis upon which subsidies in the various EEC countries were being granted. Thus whatever the amounts of money being provided by the British and French governments, the aids were premised on very different philosophies as to their purpose and the shape that they should take. The fact of intervention on an increased scale in the 1970s is not at issue, what will be discussed here is the character of that intervention. Before the two specific areas, presented at the outset, are examined, the general basis of thinking about the development of the fishing industry in Britain and France will be outlined.

French thinking has been heavily guided by the twin imperatives of national production and modernisation. There has certainly been a debate as to how best these two goals can be pursued but there has been a strong consensus that they should be pursued. This consensus has been buttressed by the way in which the alternatives facing policy-makers have been presented. In 1976, for example, the Martray report argued that France had a clear choice between maintaining national production and producers as an expression of 'a national and European maritime vocation' or buying increasing quantities of fish abroad and changing the use of the coast to other activities. The tone of the report left little doubt that the second alternative was effectively being offered as a way of reaffirming the first. Four years later in the 'Sea and Coast' document prepared for the 1981-85 Eighth Plan, there was no question of pursuing a conscious rundown of the industry.

> "France must have the specific means available to allow her to satisfy the demand for a quality product (fresh fish), at the level of production as well as of distribution."[16]

It was acknowledged that she could never be self-sufficient but it was argued that "the maintenance of an adequate level of activity at sea must play a role in limiting the deficit of the balance of trade."[17]

Though the Eighth Plan was scrapped, following the 1981 Socialist victory, the difference between the new government and the old was one of tactics, rather than strategy. The Socialist Party had underlined before the elections that fishing is "an indispensable part of the employment and regional development policies for coastal regions" and represents "an important link in our food resources", views which remained the 'leitmotif' under Le Pensec.[18] The ministry subsequently introduced a brochure entitled 'Investir à la Pêche', which underlined the state's firm support for the modernisation of the fleet and the development of production. It explained carefully the precise character of the aids that would now be available for the different parts of the industry and stressed, in particular, investment in the productive potential of new boats.

In Britain, by contrast, there was a far less developed sense of the role that fishing could or should play in the economy. Rather the concentration was upon the way in which the framework of the industry needed to be adapted for it to become economically viable. Thus modernisation of the fleet was not linked to the idea of maintaining production and cutting imports. It could only be contemplated in conjunction with a recognition of the problem of overcapacity, itself requiring a controlled running down of an artificially large fleet.

> "Surplus capacity should be trimmed away at the same time as the fleets are modernised and adapted for the future."[19]

In other words, the notion of encouraging a slimming down of the industry was officially acceptable in a way that it never was in France. By the same token, the idea of national production as a virtue in itself was never a strong one. It was no coincidence that a 1975 White Paper entitled 'Food from our own resources' failed to mention fish, annoying the BTF intensely.[20]

One good reason for this failure was that in Britain, as compared with France, there was a much more ambivalent attitude towards imports in official circles. Both in 1975 and 1980/81, there were strong demands for import controls but there was bipartisan resistance to such a move from government ministers. The Labour Under-Secretary of State for Scotland, Brown, made it clear that "a ban on imports would be against our worldwide trade relations",[21] while Walker, under similar pressure, maintained that such a ban was a complete nonsense, given the importance of maintaining supplies to the British market.[22] Though no-one in Britain would have admitted openly to supporting the second Martray alternative of boosting imports and reorientating the coast to other economic uses, the fact that trading interests were given such stress underlines the lesser British enthusiasm for support of national production as a primary goal.

It would be mistaken to infer from this that the British government was generally less inclined to intervene in the fisheries sector. In conservation matters, for example, governments of both parties were encouraged by the industry to introduce strict measures of control, limiting mesh sizes, fishing areas and particular kinds of net with a unilateral determination that provoked the ECJ in Luxembourg to condemn them on more than one occasion (cf p 92 above). By contrast, French conservation measures were less than enthusiastically received by the profession. In April 1978, for example, the CCPM immediately pressed for socio-economic assistance to help fishermen to survive whilst the resource was being reestablished.[23]

What this contrast underlines, however, is the same general concern in France about maintaining productive potential which was that much less well developed in Britain. The possibility of exclusive zones, from which foreign fishermen would be excluded, certainly helped to limit concern about the impact of restrictions on British fishermen, but even without that possibility it was widely assumed that adequate conservation measures and agreed quotas would provide the necessary framework for the industry to prosper.

> "There will be from year to year a sufficiently clear picture of the catch which will be available to the British fleets so that the catching power can in turn be tailored to suit the available stocks".[24]

For the French either the stocks would have to be related to the existing fleet or new stocks would have to be found to stop the fleet from dwindling. It was quite a different kind of philosophy.

Finally, in this section, it should be underlined that these general ideas on intervention remained substantially unaffected by the growing pressures of the period on public expenditure. In

Britain from 1976 onwards there was a decline in the level of public expenditure as a percentage of Gross Domestic Product (GDP): the era of 'cuts' had arrived.[25] In France, in the same year, Barre took over as Prime Minister from Chirac, promising a less profligate, tighter use of public money. The former made his position clear, on a visit to the fishermen of La Rochelle in June 1977: "I am not Father Christmas and I have nothing in my sack".[26] A similarly restrictive attitude to public expenditure in general emanated from the Thatcher government which took office in 1979.

Nevertheless, the actual level of financial support to the industry did not go into irreversible decline as we have seen. The editor of Fishing News congratulated Walker for squeezing so much money out of the government in 1980: "(he) can only have done a magnificent persuasive job in what must have been a hostile cabinet environment."[27] In France, even though the level of support rose substantially after the 1981 elections, there had already been a marked increase between 1977 and 1980 (80MF rising to 202MF), which was well in excess of the rate of inflation.[28] It is true that the aid for 1980 provoked terrific unrest in the industry but this was less because of its volume than because it involved an attempt to alter the basis of intervention by the state. Instead of extending the fuel subsidy which Chirac had introduced and which Le Pensec was to increase further, the government sought to introduce the idea of extra efficiency as a condition for granting aid. Its resistance to a rise in the fuel subsidy succeeded in generating a dramatic level of unity in the industry around this symbol of the productivist ethic.

As the next section indicates, such a symbol was lacking in Britain where the nature of government aid reflected a different attitude as to what sort of intervention was appropriate. How much aid would be given depended either on the skill with which individual ministers operated in a difficult economic climate or on the different political persuasions of governments. What kind of aid would be given was the product of the particular national tradition within which those ministers and governments were operating.

3. The nature of state subsidies

One of the three main changes in the fishing industry outlined in the opening chapter was the increased costs involved in pursuing the resource. What has emerged in the previous section is that the state in both Britain and France intervened to assist the industry to meet this increase in costs. However, the general philosophies of intervention operating in the two countries were reflected in the particulars of the debate over subsidies.

First of all, the two countries resorted to quite different kinds of aid. The French introduced aid to reduce the cost of fuel; in this way the expense of going to sea was automatically reduced for all fishermen. The British concentrated their subsidy on lump sum payments to owners; the amounts were calculated on the basis of vessel length, irrespective of the time spent at sea. Thus whereas in France, the subsidy was related to future activity, in Britain it was more a question of making good past losses. One Fleetwood deep-sea owner put the point as follows, when asked what he would do with his share: "all we can do is pay aid money into the bank to fill up the hole (of debt)."[29] This was simply not an option for French fishermen, who could only benefit from state support by continuing fishing.

The second important difference between the two countries was the level of continuity involved in the aids given. In Britain they provided occasional moments of relief for the industry, in France they became a permanent feature of the environment. The French began at the beginning of 1974, when the Finance Ministry agreed to a total of 20MF towards the cost of fuel.[30] Thereafter no year passed without a fuel subsidy of some kind. In 1975, the pressure of the industry was sufficient to guarantee a drop in price of between 4 and 5 centimes per litre.[31] In April 1976 Chirac proved even more generous agreeing to a rise in the subsidy for the year from 23 to 100MF, a sum worth about 15 centimes per litre.[32] After Barre became Prime Minister, the level of fuel aid was cut but it remained at 10.5 centimes per litre from 1977 to 1980.[33] What was at issue in the latter year was the industry's demand for an increase to take account of the petrol price rises of 1979. For this it had to wait until 1981, when Le

Pensec doubled the subsidy from 10.5 to 21 centimes per litre. What made him still more popular was his agreement in March 1982 that the cost of fuel should be directly linked to the cost of living, with the state paying any excess to the distributors. Here was a 'veritable lifeline' for the industry, which eliminated an important uncertainty for it.[34]

This level of continuity was not visible in Britain. Peart, the Minister of Agriculture, made no promises that the £6m allocated to the industry in March 1975 would be extended beyond the end of June and when it was extended to the end of September at a value of £2.25m, it was made clear that it was "a temporary measure to help the industry to face and adjust to changing circumstances".[35] A final £1m was found for the last quarter of the year but thereafter there were no more operating subsidies provided until 1980.[36] When the industry asked for more, the Ministry pointed out that prices were on the increase, as indeed they were for the inshore industry at least in the final part of the 1970s.[37]

When subsidies were restarted in 1980, the first package of £3m was split between support for the POs and exploratory voyages (£2 and £1m respectively). However, the payments that followed in the autumn of 1980 and then in 1981 and 1982 were like those of 1975 in that they were given to boat owners to spend as they thought fit.[38] They remained, however, *ad hoc* responses to the presssure of the industry without becoming part of a concerted government response to its difficulties. Certainly some in the industry wanted petrol subsidies: the SFF, for example, pointed out how the cost of fuel had risen in Aberdeen between January 1980 and January 1982 from 55.7p to 78.5p per gallon while the price of cod had fallen from £482 to £456 per tonne over the same period.[39] And yet such calls for more durable help found no echo in government policy.

A third aspect of the aid offered by governments to meet increased costs was the attitude of the industry towards it. The character of governmental intervention needs to be seen as much in terms of how subsidies were conceived of by the industry as in terms of the government's own policy stance. Whereas in France, there was no hesitation about turning to the state for help, the British attitude was much more reticent.

As early as 1971, when petrol prices began to rise, there was no mistaking the nature of the French industry's response: "the initial reflex of the owners was to turn to the state authorities".[40] Subsequently, when the difficulties became much more severe, it was the deep-sea owners who presented the most eloquent description of the crisis and the clearest indication that they expected the state to take the lead: "only the public powers have the authority to relaunch the maritime sector as they have done in other sectors - to protect employment and productive capacity."[41] Nor were they alone in their expectations: during the comparative calm of 1977, they kept up the pressure with the inshore industry's cooperative representatives for the maintenance of fuel aid at its 1976 level.[42] Even after a month of severe protests against the government in August 1980, it was the unanimous view of all in the CCPM, owners as well as unions, deep-sea as well as inshore representatives, that the proposals of the Minister, Le Theule, for resolving the dispute were too little and too late, not least because they failed to address the issue of rising costs by a commensurate rise in the petrol subsidy.[43] Subsequently when the new government of the Left came to power in 1981, it is noteworthy that all, whatever their political persuasion, were delighted that their expectations of help were fulfilled.[44]

The atmosphere in Britain was completely different. However much it might be argued that the subsidy offered was not enough, there was a widespread feeling that to ask for a subsidy was a last resort tactic, lacking legitimacy. When asked whether he would extend the subsidy, Peart replied: "I can make no comment. There are people in the industry who would like to see an end to the subsidy - and to stand on their own feet."[45] Nor was this ideology of independence something that the Minister imagined. After the initial aid package of 1980, one manager put the point very clearly: "the industry does not like handouts like this. We can survive on our own if we are given a fair crack of the whip."[46] The justification for aid came not from a reference to the inherent duties of the state but to the actions of other governments in giving help and thereby distorting the principles of equal competition.[47]

One consequence of this limited view of state intervention was that the issue did not escape from the party political arena, as in France. On the Left of the political spectrum there was strong suspicion about the idea of helping out the big trawler owners. In Parliament, the pressure of TGWU opposition to the owners made itself felt in Labour resistance to BTF calls for subsidies. All three of the Hull MPs (Prescott, McNamara and Johnson) argued that no public money should be provided without a corresponding public benefit, which meant in particular a challenge to the casual character of labour recruitment in the deep-sea industry.[48] By 1980, as this part of the industry was rapidly diminishing to vanishing point, Labour's attitude to aids was rather different. Strong, the Shadow spokesman on fisheries, described the August 1980 package as a "fair response", offering an "urgent and necessary lifeline".[49] The small business character of the inshore industry was not something with which the British Left easily identified but it was seen as a more worthy candidate for public money than the deep-sea companies.[50]

A second consequence of a limited state role was that governments were drawn into the invidious task of differentiating between the claims of different parts of the industry. When in 1975 all boats under 40 feet were excluded from the operating subsidy, the small-scale Cornish fishermen were furious arguing that there was a bias in favour of the big companies.[51] The same kind of bias was detected in 1980. Boats under 40 feet were included but the inshore industry pointed out that the deep-sea owners were still getting money, even though their ships were fishing for very limited periods, if at all.[52]

This problem of distribution was partly solved in France by the character of the aid: a fuel subsidy was effectively distributed to all in accordance with the amount they used. However, the industry was agreed that the effect of an increase in fuel costs was felt more greatly by the inshore than by the deep-sea industry. The latter might need more fuel but the extent of its other costs, notably crews, made energy a smaller part of its overheads. Thus in 1974 the CCPM pointed out that fuel represented about 16 to 18% of the costs of the 'artisans' and only 11-12% of those of 'industrial' companies. Moreover, the company boats made much bigger catches per litre of fuel: they needed between 1 and 2 litres to catch a kilo of fish, while the inshore vessels would use up only 0.5 to 1 litre to catch the same amount. Despite the fact that the value of the inshore species is greater, it proved possible to agree, within the CCPM framework, that the initial aid given to the industry in 1974 be handed out differentially: 6 centimes per litre for the inshore boats, 4 centimes for the deep-sea vessels.[53] Such an arrangement was out of the question in Britain: the institutions and the appropriate consensus were not there.

This section has suggested that the state aid given in the two countries to cope with increased costs differed in three important respects: the purpose for which it was intended, the degree of continuity in its provision and the attitudes of those in the industry towards it. A similarly divergent pattern can be seen if we consider the efforts of the two countries to establish or maintain fishing opportunities beyond the boundaries of NEAFC.

4. Fishing in distant waters

The spread of the EEZ doctrine in the second part of the 1970s presented Britain and France with a threat and an opportunity. On the one hand, they were faced with the prospect that their fishermen would be permanently excluded from the waters of non-EEC states that decided to apply an exclusive zone (EEZ); on the other hand, they were handed the chance to benefit from a similar zone around the island possessions that belonged to both countries throughout the world.

The creation of 200 mile EEZs posed special problems for the deep-sea industry. In the French case, this meant, in particular, the trawlers sailing to Canada and the seiners plying the coasts of Western Africa, 'la grande pêche' and 'la pêche thonière'. In the British case, it meant, above all, the distant-water fleet, with its activity concentrated around Iceland, Norway, the Soviet Union and Canada (cf Chapter Two, pp 42-3). By the end of 1976 all three categories found themselves in a very difficult situation. 'La grande pêche' which had had 30 boats in 1970 was now down to 17. Thanks to the 1972 agreement with Canada, it was assured a quota of 20,000 t.per year until 1986 but after that there was little chance of an extension, thus putting at risk 6%

of national fish production.[54] 'La pêche thonière' was taking 95% of its catch within 200 miles of the African countries between Mauretania and Angola. Its fleet of 30 ships was theoretically subject to a large degree of control by the coastal states under the new international regime, thus endangering the most profitable section of the French fishing industry.[55] As for the British deep-sea fleet, it was in even deeper trouble. Even at Grimsby, less hard-hit than Hull, there were 60 trawlers which were now excluded from Icelandic waters, and could hope for access to the other traditional non-EEC grounds only on the basis of Community-negotiated agreements.[56]

4.1 Distinctive responses

The distinctive responses of the two countries to this situation can only be understood in the context of the general attitudes towards intervention outlined above. In particular, the French enthusiasm for national production and the British concentration on national supply meant that they did not adopt similar stances. Whereas France directed its efforts towards the maximum posssible exploitation of distant waters as a way of preserving national economic independence, Britain was prepared to accept a major withdrawal from third country waters, without seeking very actively to establish an equivalent presence elsewhere.

The most obvious sign of this marked difference in emphasis was the extent of their EEZ declarations. At the beginning of 1977 both countries extended their fishery limits to 200 miles in the Atlantic in accordance with the agreement at The Hague (cf p 88 above). However France also subsequently declared a 200 mile EEZ round its overseas departments and territories, the so-called DOM-TOM.[57] To this day (July 1984) Britain has not followed the French example. The reasons for this certainly extend beyond the fishing arena. In the French case, the DOM-TOM have consistently been treated as an integral part of the French state with their own government ministry and in the case of the departments, their own representatives in the French Assembly. The British relationship with its overseas territories and possessions has been looser: they have no parliamentary representation, no separate ministry and a status distinct from mainland Britain within the EEC. Furthermore, the link between the fisheries sector and other maritime activities cannot be ignored. The very idea of extended limits was one that British official opinion resisted until late in 1976 (cf p105above). This was partly due to the conflict with Iceland but can also be attributed to Britain's traditional determination to guarantee a secure position for commercial and military shipping. By contrast, such ideas played a much less prominent role in France: the stress was on ensuring that coastal states were obliged to accept duties to accompany new rights, rather than continuing to insist on the broadest free use of the oceans.[58]

This wider background was, however, duly reflected in attitudes within the two countries to the challenges offered by the possibility of a 200 mile zone outside the NEAFC area. Even before France had formally established an EEZ around the DOM-TOM,

Giscard d'Estaing was pointing to the impact of the _de facto_ change in international law:

> "France has under her jurisdiction 11 million square kilometres of sea, which puts her in third place in the world... The sea is a new frontier of France."[59]

In this total, only 341,000 km^2 were contributed by metropolitan France and even the DOM, the overseas departments (St. Pierre et Miquelon, Martinique, Guadeloupe, Guyane and La Réunion) added no more than 668,300 km^2. The vast bulk of the President's claim, 10,172,315 km^2, came from the waters around the fourteen overseas territories or TOM (Juan de Nova, Bossas da India, Europa (all three in the Mozambique Channel), Tromelin, Nouvelle-Calédonie, Loyauté, Chesterfield, Polynésie Française, Clipperton, Crozet, Kerguelen, St.Paul et Amsterdam, Terre Adelie, and the 'collectivité territoriale' of Mayote).[60]

In Britain, by contrast, there were no equivalent ideas of establishing an extended 'maritime domain', no attempts to calculate the amount of sea to which Britain could lay claim. Indeed it was a French observer who claimed to see the full potential of the change in international law for Britain. He discussed the future in terms of the development of fishing around UK possessions, like the Falklands, and in joint ventures based on

Britain's special position in the Commonwealth, summing up as follows:

> "The 200 miles offer the United Kingdom the possibility of controlling enormous tracts of sea and of redeploying her distant water fleet over several oceans."[61]

As it transpired, the only service that deep-sea trawlers saw around British possessions outside NEAFC in the period of this study was as minesweepers in the 1982 Falklands war. Five Hull freezers (Junella, Cordella, Northella, Farnella and Pict) were requisitioned for the purpose by the Ministry of Defence and dispatched to the South Atlantic after being refitted at Rosyth.[62]

Neither country could hope to benefit from all its overseas possessions for the purposes of fishing. In only a relatively few cases is there a continental shelf of sufficient proportions to guarantee substantial stocks of fish. Nevertheless, both countries had the advantage of knowing where other countries, notably those of the Eastern bloc, had found it worthwhile to direct their efforts. Off the Kerguelens, deep in the Indian Ocean there had been a substantial effort made by the Eastern European factory ships for several years since the beginning of the 1970s. As many as 40 ships at a time were involved with an annual catch reaching as high as 250,000 t.[63] Similarly, around the Falklands and its dependency of South Georgia, there had been and has continued to be a significant effort by ships, notably from Poland, Germany, Japan and Spain, searching for hake, squid and blue whiting. The potential catch in the area was estimated at 4-5 million t. a year.[64]

The close proximity of Argentina does not make the Falklands a direct parallel with the remote Kerguelens. However, concern over Argentine feelings alone cannot explain why Britain made no move to exercise some form of jurisdiction over the waters around the Falklands. After all, France claimed a 200 mile zone around the islands of St.Pierre et Miquelon, even though it generated a considerable degree of conflict with Canada over the position of the median line between French and Canadian waters.[65] And even if Britain had not wanted to fish off the Falklands herself, she could have demanded payment of others for the right to do so. A more complete answer involves taking a closer look at the very different nature of the British state's intervention in the fishing industry in this area; it contrasts markedly with that of the state in France. Though the nature of the incentive for distant voyages was acknowledged in similar fashion, the mechanisms available to turn thoughts into action were different as were the attitudes of the participants, both on the government's and the industry's side.

The possibilities of redeploying at least some of the British deep-sea fleet were certainly recognised. The Trade and Industry

Sub-Committee specifically recommended that money should be spent on distant water trials in the South Atlantic:

> "We do not want to underrate the difficulties, but in a situation as serious as that which confronts the UK with the loss of so many traditional fishing grounds, drastic action is called for and we recommend that all concerned, both in Government and in the industry, should refuse to be daunted by the difficulties and should make an all-out effort to explore and exploit whatever resources distant waters new to them may hold."[66]

In the same year the South Atlantic Fisheries Commission urged Silkin to get the WFA to commit £1,500,000 to exploratory voyages off the Falklands to match the £6 million invested in the area by the West Germans.[67] Four years later, the Shackleton report on the Falklands called for the estabishment of a 200 mile fisheries limit around the islands, the spending of £20 million on an exploratory fishing project over five years and the extension of fishing rights to the Poles on the basis of an information exchange.[68] And yet in 1984, on a visit to South Atlantic, Battersby, MEP for Humberside, was only able to say: "The position is most encouraging .... There is scope for up to six 60ft trawlers working in inshore waters."[69] There was still no British activity in the area, despite the fact that as part of the EEC agreement of January 1983, 18 million ECUs had been made available to support state schemes for 'exploratory fishing' and 'joint ventures'.[70]

4.2 The Kerguelen experiment

The French recognition of the need to search for new grounds can be traced back as early as 1974 when the CCPM discussed the likely outcome of UNCLOS III and agreed that new grounds would have to be sought, mentioning, in particular, the Indian Ocean.[71] Hence it was hardly a surprise when FIOM was given the job, at its inception in 1976, of encouraging measures of redeployment or experimental campaigns (cf p142 above). That relatively little took place until 1979 can be attributed in part to the newness of the organisation but also to the formidable difficulties that such measures had to overcome. The main centre of interest, the Kerguelens, sometimes referred to as the 'Islands of Desolation' offer anything but an hospitable environment for fishing. They are very remote: not just 13,000 km. from mainland France but six days sailing in a southerly direction from La Réunion, the French island east of Malagasay, which acted as the last staging port for the ships involved. Lying on the 48th parallel, they are also subject to severe weather conditions. The 'roaring forties' blow for long periods of the year and there is little in the way of natural protection or shore-based, man-made installations for use in an emergency. Nor is the cost of sailing there insubstantial: one trip requires 600t. of petrol or three times the amount needed for the traditional, long-distance voyages in the Atlantic to Canada.[72]

Despite these difficulties, in 1979 one of the firms engaged in fishing off Canada, the Société Nouvelle des Pêches Lointaines (SNPL) did organise a trial off the Kerguelens with a 5MF grant from FIOM. In August of that year, the freezer trawler 'Jutland' left its home port of Bordeaux and after an initial fishing sortie off Africa, made its way to the Kerguelens, arriving in October and staying for four weeks. From one point of view, the voyage confirmed Le Monde's verdict of it as "rather disappointing".[73] Only 78t.of fish were caught, suggesting that the voyage took place at the wrong time of year, when fish were scarce and the weather deteriorating. Nevertheless, from another point of view the very fact that the attempt had been made and that FIOM had provided financial support, showed that it was no ordinary venture:

> "Whatever else, the experience of the Jutland has a somewhat symbolic aspect. In fact it testifies to the will of the French fishing industry to face up to the limitations on access to the resource. It also marks the will of the public authorities to assist the initiatives taken by the profession."[74]

Certainly there was no slackening of interest on either side. In 1980 FIOM reserved over a sixth of its total budget (22.5MF) for distant-water trials, while the navy expressed its willingness in principle to extend the help it offered off Canada to the depths of the Indian Ocean.[75] In the spring of 1981, fishing companies from Boulogne and La Réunion combined to form PROMERSUD with the specific aim of developing fishing capacity in the Southern hemisphere "to guarantee the independence of the French fishing industry."[76] PROMERSUD then based two ships, the

'Austral' and the 'Sydéro', at La Réunion and from there they made one and two voyages respectively during the summer of 1981. As for SNPL, she maintained her interest by sending out the 'Zélande' to fish during June and July. This time the result was much more successful. The 'Zélande' caught nearly ten times as much as its predecessor, unloading 700t. on its return to Bordeaux in August. The atmosphere was that much brighter that now the talk was of "the Kerguelen treasure."[77] All the owners were starting to establish a much clearer picture of what to catch, when and where, given that the three main species concerned are abundant at different times of the year.

Once again, however, the importance of state finance channelled through FIOM cannot be over-emphasized. It is most unlikely that even with the better prospects for 1981, the owners would have taken the risks involved without the promise of financial aid. In the event, not all the 15MF available was claimed but the reassuring thought that such sums would continue was important in their calculations. Dezeustre, the Head of SNPL, saw financial aid from the state as a long-term necessity: "the pump must remain primed for the day when the whole fleet (of Canadian freezers) is deployed around the Kerguelens".[78] In 1982 the state continued its support and the 'Zélande' set off again in April for the Indian Ocean suggesting that the trips were becoming routine.

However, state aid by itself was not enough. Despite the active efforts of FIOM to market the product, it proved very difficult to find buyers anywhere in the world for the unusual species on offer. By the end of 1982, SNPL was saying it would not return to the Kerguelens in the following year and was worried even about continued access to Canadian waters, following the EEC ban on the import of baby seal skins.[79] By the spring of 1983, it was not the companies but the state that was maintaining an interest in the islands. The French navy bought one of the freezer trawlers, the 'Nevé', and proposed to convert it so as to be able to keep an eye on the Russian trawlers still operating in the area.[80] The 1981 prediction that by the middle of the decade annual French production would reach 40-50,000t. at a value of 100 to 125MF guaranteeing the viability of the nine remaining ships of 'la grande pêche', looked decidedly unlikely.[81]

4.3 The tuna fleet

The difficulties that the Kerguelen trials met should not conceal the determination that was shown in France to take advantage of the new 200 mile zones. It was a concerted strategy to maintain national production which was by no means restricted to this one case. The tuna fleet also received important state support, both via FIOM and through direct diplomatic backing. As a result, the owners were looking forward in 1982 to an important level of expansion: over 30 ships in 1987 compared to 23 in 1981 and a catch of 55,000t. rising to 90,000t. over the same period.[82]

Although no-one saw this part of the deep-water fleet as being as severely threatened as the freezers of the Northern waters by the creation of 200 mile EEZs, there was a recognition that it could not stand still. The companies themselves started modifying the catching areas of their boats, so that by 1983 around 40% of their catch was being taken outside the 200 mile zones of the African states, compared to only 3% in 1973.[83] However, this still meant that the bulk of their activity was within the zones and there was increasing pressure from at least some African states, like Senegal, for the French to pay for the privilege of fishing.

The possibilities of fishing further afield, in the Pacific, were discussed in 1979 at a meeting in New Caledonia, attended by the Secretary of State for the DOM-TOM and the Chairman of the Interministerial Mission on the Sea. It was suggested that 15,000t. could be caught off New Caledonia and as much as 40,000t. off French Polynesia. The owners argued that the distances and costs involved were excessive but the idea of redeployment was on the agenda.[84] The industry set up a consortium called COFREPECHE at the beginning of 1981 with the explicit intention of diversifying the activities of the tuna fleet. The first overt evidence of its activities was an experimental voyage at the end of the year in the Indian Ocean by the 'Yves de Kerguelen', which was able to have 580t. of fish shipped back to France.[85]

Here too a close relationship developed between the state and the industry. In financial terms, FIOM provided 8MF out of the 14MF that the voyage cost, while scientific help came from the <u>Office for Overseas Technical and Scientific Research</u> (ORSTOM), a government sponsored body with particular responsibility for research in the intertropical zones.[86] The arrangement was repeated at the end of 1982, when four ships sailed and FIOM again provided 1.6MF to cover any losses, money without which the owners maintained that they would not have gone ahead.[87]

However, it was not simply a question of financial and technical support: there was no shortage of diplomatic effort to guarantee maximum access to the main grounds off western Africa. Despite the formal change in 1977 to Community-negotiated agreements with third countries and the interest of other EEC states in these grounds, notably Italy and Greece, France did not give up all bilateral contacts with African states. Le Pensec, for example, invited the President of Equatorial Guinea to Concarneau in the autumn of 1982, after the latter had banned tuna fishing off his coasts the previous year. They came to an agreement by linking the re-establishment of fishing rights to the acceptance by the tuna fleet to supply salt to an impoverished island off the Guinean coast.[88] After a little local difficulty - the ship delivering the salt was arrested - a formal accord between the EEC and Equatorial Guinea was subsequently reached in June 1983. This supplemented Community agreements that had already been made with Senegal, Guinea-Bissau and Guinea (Conakry).[89]

4.4 Contrasting policy styles

It would be naive to make a direct comparison between France and Britain and to chide the latter for failing to support her deep-sea industry as the French did. For one thing, there was no tradition of fishing in tropical waters or indeed anywhere outside the cold Northern waters; for another, bilateral deals with the coastal states familiar to the British industry were immensely complicated not simply by EEC membership but by the number of other supplicants and the strength of the domestic fishing industries of the coastal states concerned. Canada and Norway are a very different proposition from Senegal and Equatorial Guinea. Rather it is necessary to seek an explanation for Britain's failure to act like France in the kinds of attitudes that both government and industry had about each other's role. However close individuals from the industry might come to the policy process, there remained a strong conviction on the part of government that it was up to the industry to manage its economic future and on the industry's part that it should be left to get on with making a living.

The different British policy style was illustrated in the exploratory voyages that were undertaken by or on behalf of the industry. Already before the clash with Iceland major efforts were made to search for alternative species for the deep-sea industry. Excitement was considerable about the possibilities of blue whiting, which was known to exist in large volumes near Rockall. In March 1975, for example, MAFF and the WFA coordinated to hire the Hull trawler 'Arctic Privateer' for a 50 day voyage which proved very successful with a catch of 400t.[90] However, there was all the difference in the world between this kind of experiment and the owners themselves deciding to go in search of the blue whiting. One Fleetwood fishing manager pointed to the price of £30 per ton in the auction and insisted that it was not worth his company's while: "we're not desparate".[91] From a market-orientated position, his position was completely rational: after all, what chance did the blue whiting have when Birds Eye were spending £10m per year on encouraging people to eat cod fillets?[92]

The only way that such logic could be modified would be through conscious state attempts to manipulate the market, as FIOM was effectively doing in France. However, there was no question of this in the British context even after the deep-sea fleet had started to contract dramatically. In January 1980, for example, the government put up £443,000 to finance four freezers to look for scad (horse mackerel) in the Bay of Biscay and placed scientists on board each ship to monitor what happened. At the same time, it made it clear that all responsibility for marketing the catch was the owners'. Hence the fact that the value of a large catch was undermined by the import surge of that year was seen as an unavoidable accident, not a condition to be combatted.[93]

The aid that the government subsequently gave in March 1980 followed in this 'non-directive' tradition. One million pounds were made available for exploratory voyages but the money was declared 'open for tender' with anyone free to apply and no attempt made to determine with the industry as a whole where and what was to be fished, other than specifying 'under-utilised species'. By the time the next package of aid came along, all further exploratory voyages had been cancelled because there was no more money and no market for any fish caught was perceived to exist.[94]

Within government there was a tendency to argue that much of the problem lay with the owners themselves. When questioned by the Trade and Industry Sub-Committee about why there had not been more attention on distant waters outside NEAFC, Kelsey, the Fisheries Secretary at MAFF, replied: "I am bound to say ... that the British industry has not shown much interest in these more widespread, long-ranging possibilities".[95] However, to say this was to acknowledge that it was not considered appropriate to exercise undue influence over the industry's perceptions of the possible. Nor was it entirely accurate in that there were attempts to escape the growing level of inactivity but none of them enjoyed the kind of supervision exercised by FIOM.

One early independent initiative came in 1977 when the Boston Group and J.Marr sent three and two ships respectively on trials in the waters between Greenland and Canada. The results were disastrous with large quantities of the catch not sold and both firms suffering heavy losses: £60,000 in the case of the Boston Group; £29,000 in the case of J.Marr. As a consequence, three more trawlers were laid up.[96] Then in 1978/9, three British United Trawlers (BUT) vessels, the 'Othello', 'Cassio' and 'Orsis' were sent to fish off Western Australia. BUT had taken a 50% share in an Australian company, Southern Ocean Fish Processors Ltd., and wanted to develop a catching and processing centre in

Albany. However, the company found that the substantial mix of species in these sub-tropical waters made sorting and grading difficult and resulted in too small a catch of commercially-viable fish. The Australians, for their part, thought that BUT had not done enough on the marketing side and so the experiment came to an abrupt end.[97]

What both these examples underline is that within Britain, the industry generally preferred to go its own way and the state had no institutional mechanism by which to channel the energies of the owners, even if a desire to do so had existed. FIOM, by contrast, was the expresson of the French industry's willingness to have its preferences manipulated and of the state's intent to engage in such manipulation. It is true that FIOM's role did arouse controversy: for some time after its birth, the owners looked upon it with suspicion as yet another bureaucratic block to progress. In one notable article, the UAP took a phrase from a popular analysis of the basic weakness of French society, to draw attention to what it saw as wrong with the workings of FIOM.

> "The technocrat's dream is to fix the future on paper, coldly deciding what is suitable for the administered." [98]

However, it then emerged that the owners' main complaint was not that FIOM was doing too little but rather that it was not giving enough backing for distant water fishing.[99] In an important sense, the owners did not want to go their own way.

It does not follow from this that there was automatic agreement in France on all issues. There was certainly not unanimity, for example, about the financing of the Kerguelen voyages. When the discussion for the 1981-3 period was underway, 5 members of the Management Council abstained in the vote, doubtful whether it was worth spending the money and feeling that the other main company with a deep-sea interest was being discriminated against.[100] What FIOM permitted, however, was the possibility of such sectional squabbles being overcome.

A similar process of interest aggregation was visible in the way the demands of the unions were treated. The leaders recognized the difficulties that they were obliging their men to face around the Kerguelens. The sailors would be forced to stand up to very severe weather conditions with poor communications and little assistance available. And given the lower wage rates applicable in La Réunion, they could not be sure that in the longer term, they would keep their jobs. Furthermore, their present earnings were at risk in that they could not know whether the 20% of the value of fish disembarked that the crew received would bring them their normal level of earnings.[101]

As a result, the CFDT, for example, resolutely opposed voyages exceeding 135 days and demanded that state support be provided to pay for reserve crews to be flown out.[102] Such demands were generally regarded as reflecting a recognition by the unions (and that includes the usually uncompromising CGT) that they needed to avoid excessive demands, if the trials were to have any chance of succeeding.[103] Indeed they achieved a partial satisfaction for those demands. Though not realised, the plans for 1983 included an agreement that the crew would be sent home by air for 15 days between each of the three 75 day voyages that the 'Zélande' was to make from La Réunion.[104]

Such moderation on the part of the union and the agreement of the owners can be seen as the product of the recognition that there was no alternative. Yet the perception of alternatives is a relative thing. However much individuals or committees might encourage distant water voyages in Britain, nothing was done, and this inactivity was for reasons that might equally well have been used in France. The men were said not to be prepared to go and be away from their families for such long periods of time, while the owners complained that the nearest available dry dock facilities in South Africa were far too distant.[105]

It is clear that the logic of the policy process worked in very different ways in the two states. In France, there was plenty of hard bargaining, with the owners arguing that it was really all too expensive, the government maintaining that it could not possibly cover all the risks and the unions insisting that the conditions for the men on the ships were intolerable. Nevertheless, these stances formed part of a ritual where "everybody played the game"[106] and where a common committment to boosting national production ensured that an agreement would eventually be reached. In Britain, by contrast, there was not this same underlying consensus about what the purpose of such trips would be and hence the disagreements between the partners served as a force for fragmentation, with the state not able or willing to act to aggregate the various interests involved.

5. Conclusion

This chapter has indicated "that the industry would demand increased intervention to protect it from the effects of the changing environment and that the government would feel itself obliged to cater for these demands". However, it has been found that the extent of the intervention demanded and the nature of that intervention varied greatly in the two states. The evidence confirms our assumption "that the 'dirigiste' French tradition and the British aversion to intervention would mean that government efforts to determine the shape of the industry were more developed in France than in Britain".

The material has not suggested that the French state, unlike its British counterpart, will always act innovatively or impose a central direction. Rather it confims the view that there is in the former:

> "a capacity for policy initiative, a potential for far-sighted planning and a propensity to impose its will when this is necessary to obtain public objectives."[107]

This readiness to use such a political will seems to be particularly relevant in the context of the economic objectives of national production and modernisation, where a broad consensus exists anyway and to which the state can appeal with powerful effect. In Britain, where the concept of market supply is more powerful than that of production, such an appeal is not available to overcome divergent private objectives. Instead the British state responded to calls for intervention in a way which reflected a limited desire to establish a clear ordering of priorities within the fisheries sector: the public interest was no more than the sum of individual interests.

Finally, the chapter has not provided an easy answer to the issue of the comparative success enjoyed by the two fishing industries. The breakdown of the Kerguelen experiment underlines the difficulties involved in seeking to manage the fish market and to find international outlets for national production. Similarly, it is not obvious that any more widespread fishing by British deep-sea trawlers in distant waters would have had a major impact on their declining fortunes. Rather the chapter shows how the two states pursued different strategies to cope with decline, which operated on the basis of very different premises as to what both the ends and means of policy should be. Judgement on the success of the industries cannot therefore be separated from a judgement of the wider merits of more liberal as against more protectionist economic philosophies. We shall return to this issue in the final chapter.

6 The mediatory perspective

1. Introduction

This chapter will broaden the discussion of the relationship between state and society in the fisheries sector, by considering the importance of the actors outside the immediate circle provided by the government and the industry. As in the previous chapter, the material will revolve around the two assumptions linked to the perspective, which were introduced at the beginning of this study (p 30 above). It was suggested, first, that "the representatives of the industry would be obliged to transmit the views of the industry through a much wider set of channels because of the disgruntlement of their members at the effects of the changing environment". The second suggestion was "that higher French acceptance of the legitimacy of public action and the greater elasticity of British political life would make such developments more widespread in Britain than in France." Here it will be claimed that the broader ventilation of discontent was a central aspect of the behaviour of the British industry. However, the French industry showed little interest in actively pursuing the support of actors outside the traditional institutional framework. This challenge to the first assumption will be explained in terms of the importance of the parliamentary arena in the British context and its relative weakness in the French setting. The same contrast also helps to explain why private interests within the industry were able to flourish in Britain in a way that they could not in France, thus underlining the validity of the distinction contained in the second assumption.

To develop the argument, the chapter is divided into two parts. In the first, it will be claimed that the political salience of the issue increased in both countries but in different ways. Whereas in Britain, the industry consciously sought to influence government through Parliament by regular appeals to a wider opinion, outside interest in the fisheries issue in France was much more sporadic and developed without direct pressure from the industry. The second section will suggest that these contrasting patterns in the mediation of grievances had different effects on the relative cohesion of the two industries. In Britain, divisions within the industry generated strong centrifugal pressures, with the state's role being limited to that of a broker, reliant on consultation and consent. In France, a centripetal process was at work. The importance of the state's role in aggregating individual interests meant that separate groups within the industry were restrained from seeking preferential treatment vis-à-vis other groups by canvassing for support in the political arena.

The chapter should be seen as closely linked to the previous one. It develops the idea of a tension between the aggregation and the articulation of societal interests and the contrasting ways in which that tension was resolved in Britain and France. Just as the previous chapter concentrated on the French system with its capacities for aggregation, so the present chapter will lean more towards the British experience with its stress upon the fullest articulation of interests in the relations between state and society.

2. Divergent patterns of political salience

One of the paradoxes of this study is the gap between the economic and political importance of the fishing industry. Whatever the industry's difficulties, it was and remains, in a relative sense, of minute economic importance. In none of the present members of the Community do fishermen represent more than 1% of the total workforce, in most the figure is much lower. In 1979 it was estimated that the percentage in Britain was 0.09 and in France 0.11.[1] Similarly, the contribution of landings to the Gross Domestic Product (GDP) of the present EEC states is well under 1% in every case; in 1976 it was put at 0.17% in both France and Britain.[2] A comparison with the agricultural sector underlines the point. In 1977, for example, there were over 2 million people employed in agriculture in France, just under 10% of the workforce, contributing nearly 5% of GDP; in Britain in the same year there were many fewer - just under three quarters of a million - but that still represented 2.8% of the workforce and the same percentage of GDP.[3]

At the same time, the fisheries issue assumed a much greater political importance in both countries. The change was epitomised by Foreign Secretary Crosland's comment that British foreign policy had been reduced to "fish and bloody Rhodesia", while in France, the issue reached a highpoint of resonance when it figured in the debate between Mitterand and Giscard d'Estaing before the 1981 Presidential election, the former accusing the latter of a 'lack of tenacity' in the CFP negotiations.[4] Such comments would have been unthinkable even five years earlier. The French Council of Ministers only had its first overview of the problems of the sea at its meeting on December 15 1976, while the debate on fisheries in the British Parliament on June 30 1975 was the first to be called for by the Opposition front bench rather than individual backbenchers.

However, the extent of this new political salience and the processes through which it was brought about were very different in the two countries. Whereas the issue rarely disappeared from view in Britain, it was only in 1975 and then again in 1980-1 that the case of the French fishermen emerged from obscurity. This contrast was in part due to the two industries being surrounded by very different political environments and in part the result of their conceiving of their relationship with the state in divergent ways.

It has already been shown in Chapter Four how the fishery sector's relationship with the state had developed very differently in the two countries. From the time of Colbert there had been in France a close identification between the two, maintained through a set of state-sponsored institutions. In

Britain the ties had been much looser, with little attempt to mark out fishermen as a group from the rest of the population. These differences were to be one reason why in France the industry did not actively seek to ventilate its grievances through a wider set of channels, whereas in Britain the opportunity existed and was eagerly grasped by the industry.

However, there was a further reason for the different levels of political salience which relates to the contrasting political arrangements in the two countries. The Fifth Republic's Constitution produced in 1958 was a reaction against what were seen as the deficiencies of its predecessor, in particular the role of political parties in undermining the stability of government through the Assembly.[5] This assault on the 'gouvernement d'assemblée' resulted in a clear institutional separation of the executive and the legislature combined with a marked downgrading of the powers of the legislature. In Britain, by contrast, whatever the actual limits on Parliament's power, the political system continued to revolve around the principle of the sovereignty of Parliament with the executive integrated into and working through the legislature. Furthermore, the long-running coherence and discipline of the British two party system was unmatched in the French context. The 'revitalisation' of the parties in France in the 1970s was a matter for surprised comment, rather than a basic feature of political life.[6]

These two elements help to explain what happened when the difficulties of the fishing industry grew in the 1970s. In France, as we have seen (cf p 141 above) the feeling spread that the CCPM structure was inadequate as a way of protecting the economic interests of the industry. Hence the creation of FIOM at the beginning of 1976 and the continuous pressure for a Ministry of the Sea. However, it is important not to misunderstand the character of this dissatisfaction. There was still a very strong residue of feeling that the old structure remained the central element for transmitting the views of the industry to the state. Never was this feeling more evident than when outsiders sought to challenge that structure. In 1981 the head of a small independent research body, dealing with fishery matters, the Centre d'Etude et d'Action Sociales Maritimes (CEASM) wrote a scathing attack on the organisation of the maritime sector:

> "Attached to the Transport Ministry between Concorde and the Airbus, the maritime profession are, in the eyes of the administration that supervises them, sailors rather than producers. The CCPM, born under Vichy, provides the sector with a corporatist professional organisation. The unions are up to their necks in it, stuck between the owners, the merchants, the fishmongers and the administration."[7]

The response was immediate and powerful. Not only did the CCPM's President deny the link with Vichy but its Bureau, with representatives from all sides of the industry, passed a unanimous resolution disapproving of the attack and reaffirming the value of the 1945 institutions.[8]

This restatement of faith did not mean that the CCPM was considered, even by those within it, as the only channel of communication with the government. The representatives of the companies, in particular, stressed the importance of direct links with the Ministry and regarded the CCPM as a safety valve rather than as a way of influencing government policy. For them it has remained a structure which helps to counterbalance the privileged access of the employers to the Ministry but which can do no more than allow the administration to hear the industry's response to decisions taken elsewhere.[9] The unions, for their part, were well aware of the limited role of the CCPM, notably in the exclusion of discussion on issues where there was a major lack of consensus. As one of their leaders put it, "life would be intolerable" if issues such as the size of crews or the level of salaries were raised in the central or local committees. Such issues had to be dealt with directly with the employers while support for the union view was sought directly from the government, particularly after the 1981 elections.[10]

However, these reservations were not new ones, nor were they substantially affected by the changes of the 1970s and 1980s. The owners did not consider boycotting the 1945 institutions, however much they might question privately its overall importance. The unions also continued to see the benefits of a formal, legally-defined set of arrangements, which provided them with a guaranteed means of making their views known to government. For their different reasons, neither side had a strong incentive to look beyond the existing framework as a way of exerting more pressure on the state apparatus.

In Britain, by contrast, the pattern of relations saw considerable development in the 1970s. The industry devoted an increasing amount of time and money to bringing indirect pressure on the government. Far from stressing direct contacts, like their French counterparts, British fishermen sought to influence policy by obtaining maximum exposure for their grievances outside the narrow confines of the Whitehall ministries. Above all, the parliamentary arena became the focus of the fishermen's defence of their interests in a way which had no parallel in France.

There had always been close relations between the fishing industry and individual constituency MPs. For many years these MPs had met within the framework of the All Party Fisheries Group to discuss the difficulties of the industry and had used debates in the Commons to prod the government into action. In 1971, at the time of negotiation on accession to the EEC, it was the 20 to 30 Conservative members representing the constituencies of the inshore industry who had made the most fuss, pressing for a revision of the CFP agreed by the Six the year before.[11] Similarly, during the three 'Cod Wars' (1958-61, 1972-3 and 1975-6), it was the mainly Labour members of the deep-sea ports of Hull, Grimsby and Aberdeen who had spoken most frequently on the floor of the House to bolster anti-Icelandic opinion and to try to stop the government from any back-sliding.[12]

There was therefore already a basic framework of contacts between the industry and Parliament, upon which it was possible to build. This was in marked contrast with the situation in France, where attitudes to the members of the Assembly had never been particularly positive. With one or two exceptions, notably Le Pensec, a deputy for many years in Finistère before he became Minister for the Sea in 1981, the bulk of deputies were considered to lack "le fibre maritime" i.e either major feeling for or expertise in the issue at stake.[13] The British industry, for its part was unanimous in believing in the importance of Parliament as a way of influencing governments. Though the SFF and TGWU would probably agree on little else, their Chief Executive and National Fishing Officer respectively gave very similar appreciations. The former described Parliament as 'the best road', while the latter called it 'the best tool', perhaps the only tool when an individual or group did not have the ear of a minister.[14] Thus the activities of the industry in the period between 1975 and 1983 remained predominantly within the parliamentary framework.

However, the range of strategies used by British fishermen extended far beyond anything that had been known in the past. Both at Westminster itself and beyond they devised techniques to make life as uncomfortable as possible for any minister who was prepared to sacrifice this sectional group in the interests of some wider objective. Lobbying expanded from a port-based activity onto a much wider scale, reflecting the increase in organisational coherence of the industry. The development of the SFF, for example, made it possible to coordinate the descent of fishermen from all over Scotland onto Westminster in February, 1979 to keep up pressure on the government. As the newsletter of the SFF put it: "the delegation lobbied MPs of all parties. MPs and passers-by were given parcels of haddock and told 'have some while there's still some left'."[15] Similarly, over 300 fishermen from the South West came up to London in March 1981, fearful that an EEC agreement might fail to protect them adequately. "Exclusive limit or we're on the dole" was the cry.[16] Again such concerted action between ports would have been quite out of the question in earlier years.

Within Parliament this kind of pressure was reflected in the increased volume of debate and the care that the government took to placate the fisheries lobby. Following the major debate of June 30, 1975, the issue was rarely out of the pages of Hansard for more than a few months.[17] In particular, every time there was an EEC Council meeting, Ministers would make an appearance in the Commons before, after or both. If it was before the meeting, they could follow the fullest airing of the industry's case with a promise that they would not ignore what had been said. As Silkin put it in June 1978:

> "I shall go to Luxembourg on Monday in the knowledge that the House is behind me in demanding a fair deal for British fishermen."[18]

On their return, Ministers would be able to explain how such a fair deal had eluded them or as occurred in January 1983, how success had been achieved. Even in this latter case, though, Walker was mindful that he was not just talking to the assembled MPs:

> "I would like to record my gratitude to the leaders of the fishing industry who have attended every meeting with me and who have discussed and agreed what we have negotiated."[19]

Such overt use of the French Assembly as a medium for the grievances of the industry was out of the question. Only once a year was the fisheries issue even formally discussed and that was in the context of a debate on the budget of the Merchant Marine as a whole. On one occasion in June 1977 there was a specific debate on the sea but it aroused little enthusiasm. The government had already put it off several times and attendance was poor. It seemed difficult to believe that 150 people claimed to belong to the Parliamentary Group for the Sea.[20] In general, the members within the Assembly were seen as not very concerned with the issue. The uproar of August 1980, when suddenly everyone was active in the fishermen's cause, was contrasted with the annual Merchant Marine debates where "the emptiness of the chamber is broken only by a dozen deputies or senators struggling against the sandman."[21]

What made the situation so different in Britain was the existence of two dominant and relatively well-disciplined parties, alternating in office. The industry was aware that getting its view accepted in Parliament meant obtaining influence over the present government or the alternative to it. In fact it was remarkably successful in extending support for its position over both parties, generating a high degree of consensus over the need to protect fishermen, particularly in the context of the EEC negotiations. Within debates the result was that nearly everyone seemed to be saying pretty much the same thing. As Hamish Watt commented: "There is such a degree of unanimity on this question that it is difficult to avoid repetition."[22] It also meant that there was an awareness that going outside the consensus might be expensive electorally. In 1978, for example, it was calculated that among the 22 fishing constituencies, the percentage of marginal seats was three times the national average with the Conservatives in particular vulnerable to any swing.[23]

The industry, for its part, was careful not to upset this consensus by lining up with any of the minority parties. This was particularly noticeable in Scotland where the Scottish Nationalist Party (SNP) took up the fishing issue with great vigour: indeed it was very much thanks to its member, Hamish Watt, that the Trade and Industry Sub-Committee's enquiry into the fishing industry was set up. However, the Scottish fishermen's representatives were careful not to get too closely involved with the SNP. As Hay, the President of the SFF, put it:

> "We are grateful for the interest and help the Scottish Nationalists have shown in our industry but we won't be used as a political tool."[24]

Thus when in 1979 the SNP lost the four Grampian fishing constituencies of East Aberdeenshire, Banffshire, Moray and Nairn and South Angus, it was not difficult for the SFF to carry on its efforts through the new Conservative members.[25]

The party structure in France was quite different and did not offer the French industry the same opportunities, even if they had wished to take advantage of them. It was not that they were totally unaware of the grievances of the fishermen but rather that like the SNP, they sought to use those grievances to develop their own overall political profile in the eyes of the electorate. This was particularly true of the two parties of the Left, who were in opposition until 1981, the Socialists (PSF) and the Communists (PCF). Both set out to establish a distinctive policy approach for an issue that they had not given much, if any, thought to in earlier years. For the PCF, this meant tabling draft laws in October 1976 and May 1980, which were based on themes familiar from other parts of the party's programme, such as the inadequacy of state support and the increasing grip of the multinational trusts upon French production.[26] For the PSF, a first study day on fisheries in 1976 was followed by the development of a number of policy proposals which were put together into a book for the

1981 elections.[27] La Mer Retrouvée reflected the wider thinking of the party with its stress on the need for investment and for a more developed social policy. In this way both parties were able to confront their adversaries in the 1981 election with a set of distinctive proposals for this issue and to give a political dimension to the fisheries' issue that it had not had before. Indeed it appeared to be a notably successful strategy in combatting Giscard d'Estaing. When the second round of the Presidential elections took place in May 1981, it emerged that in Brittany in particular, the fishing ports turned markedly against the outgoing President.[28] Once the elections were over, however, the issue returned very much to the obscurity from which the political parties had raised it.

In Britain, by contrast, the issue retained much greater salience throughout the period of this study. This was because the industry's efforts extended beyond the direct links that were forged with individual MPs and political parties. The use of lobbying at Westminster was by no means the only way in which the catching sector sought to influence opinion there. What occurred was a proliferation of contacts, all designed to keep the pressure on and make life difficult for the government. It was a development which knew no parallel in France.

The most obvious indicator of the change was the way in which the media became a central element in the defence of the industry. The wealthier sections of the industry, notably the BTF, had traditionally employed a public relations consultant to present their case but now an effort was made to coordinate the industry's position. On July 20 1976 the inshore and deep-sea industries launched an advertising campaign in the national press at a cost of £35,000, urging their case for exclusive limits of at least 50 miles.[29] Therafter it was never difficult for the industry to obtain a hearing, whether in the newspapers on televison or on the radio. The voice of Austen Laing, for example, was regularly heard on the 'Today' programme, commenting from Hull on the latest failure to achieve an EEC agreement.

Moreover, there was no problem in using the media to get across a particular version of events. Reference has already been made (p 93 above) to the March 1980 'World in Action' programme when a Hull trawler skipper claimed to have found illegally caught herring on the quayside in Boulogne. A conspicuous feature of the programme was that the viewer was hardly given a chance to assess the validity of the charge. What he was given was the opportunity to share in and subscribe to the intense suspicion felt by the British fishing industry towards the industry in other countries.

This vision of the world also made its way into the parliamentary arena via the press. When Fishing News reported similar illegal landings of herring later in the year, it was immediately picked up by MPs in the Commons. Johnson, one of the Labour MPs for Hull, waved his copy of the paper furiously at the government and demanded action against the culprits, the French.[30] Nor should it be supposed that what appeared in such a paper was treated as of no account by government ministers. Buchanan-Smith,

Minister of State at MAFF after 1979, accused Fishing News of disreputable journalism when it put out a banner headline: "We must give way" i.e in the EEC negotiations. While he denied having said anything of the kind, the paper retorted that this was certainly the tenor of his remarks.[31] There was an implicit recognition on both sides that what the paper had to say was influential and could make life very difficult for the government.

The contrast with France could not be more marked. The sister paper of Fishing News, Le Marin, often mirrored the British paper's coverage but it remained very much a trade journal, without any wider political significance. After reporting the herring catch story from Fishing News, it did turn to the offensive itself. It pointed to the way in which Spanish vessels were registering in Britain and fishing under the British flag. The claim was that this was a deliberate tactic because there were not enough native ships to catch the level of quotas that Britain was demanding in the EEC negotiations.[32] Within the French industry this story provoked considerable uproar but there was no wider arena domestically into which it could be projected for the purpose of exerting influence. The French Parliament simply did not offer an arena for the ventilation of grievances in the way that Westminster did.

More generally, the link between the media and the industry in France remained very underdeveloped. There were certainly plenty of stories about the hardship of the fishermen's life and at times, long articles on why the industry was in the state it was, but the relationship remained a distant one.[33] Even with its base in Paris, the UAP was lucky to get one request a year from radio and television to present its view.[34] Nor was it necessarily easy to persuade the media to take a greater interest. Dubreuil, President of the CCPM, tried to get the television channels interested in a programme concerning the effects of the ban on the import of baby seal skins on the fishermen of 'la grande pêche'. They were simply not interested and he did not pursue the matter.[35]

In Britain nothing seemed easier than to attract media attention and the fishermen proved markedly adept at it. They developed a further tactic to increase their salience in the public eye, which was unmatched by the fishermen of any other EEC country. Though a few members of the French fishing industry had traditionally gone to Brussels to give their delegation support, it was nothing compared with the trips that the British industry made every time the Council of Fishery Ministers met. From 1977 onwards, from 15 to 40 representatives would fly off to Brussels or Luxembourg and stay there for the duration of the meeting. To some this looked like an enormous amount of time, effort and money for very little reward, when the important consultations had already taken place in London before the meeting.[36] It can be seen, however, as much more of a public relations exercise, designed to convince the men back home that a good job was being done with their subscriptions and to make sure that any concessions in the Council chamber got as much ventilation as possible after the meeting. The Minister might have his own version of events but the fishermen's presence ensured that the journalists could be guaranteed an immediate industry quote on the official version.

This final example of the strategy of British fishermen underlines again how the nature of their relations with the state and the character of the political system as a whole encouraged behaviour which was quite out of the question in France. It should, however, be said that the resonance of the issue was accentuated by other factors, which made it a rather special one in British eyes. For one thing, whereas France has never had a highly-developed 'esprit maritime', for Britain the sea has always had a very strong symbolic content. Canetti has pointed to the way in which all the triumphs and disasters of English history are bound up with the sea, and suggested that it remains a distinctive 'crowd symbol':

> "The Englishmen sees himself as a captain on board a ship with a small group of people, the sea around and beneath him. He is almost alone; as captain he is in many ways isolated even from his crew."[37]

Though this link with the sea necessarily remains difficult to quantify, there can be little doubt that it helped to generate a strong sense of admiration for the fishing community. Once that community was under threat, support for it assumed an emotional aspect which it could never have in the French context with traditions much more closely associated with the land.

There is also a second sense in which the nature of the issue was a rather special one for Britain. It cannot be forgotten that for most people in the British fishing industry a basic problem of the CFP was that it had been agreed in June 1970 on the day before the negotiations with the applicant states began: for them it lacked legitimacy. France, by contrast, had put her name to that agreement and subsequently insisted that it was and must be the legal starting point for any further discussion. Hence the British were seeking to challenge the whole basis of the debate on the CFP, while the French worked to restrict negotiation to matters which did not put that basis in doubt. Britain's isolation in the EEC and the bias in the policy process generated by the attitude of the other member states help further to explain the increased politicisation of the issue.[38]

These arguments do not, however, run counter to the main contention of this section. Rather they offer a wider setting for the basic distinction offered between an intermittent salience in France to which the industry contributed little, and a regular salience in Britain which the industry did all it could to develop. The effects of this distinction will be considered in the next section.

3. The relative cohesion of the two industries

In the previous section the strategies of the industry were discussed in unitary terms. However, it should be clear from Chapters Two and Four that the industry in both countries is a very varied one. Leaving aside the division between the catching sector and those outside it, there are significant distinctions amongst the catchers between inshore and deep-sea interests, employer and employee interests and different regional interests. It is these distinctions which will be considered in turn in this section. In France they were contained reasonably well within the existing structure of consultation, whereas in Britain the greater ventilation of the issue served not only to increase its general political salience but also to underline the divergent priorities of the various sections of the industry.

It will also be argued that the mediatory process cannot be understood without reference to the role of the state and the issues raised in the previous chapter. The nature of the economic intervention that was considered legitimate both by the state and by the industry was itself important in determining the way in which demands were formulated, pursued and resolved within the political arena. The greater British reluctance to prognosticate on the desirable economic shape of the industry left wider scope for debate and offered less common ground around which all fishermen could coalesce. There was therefore a centrifugal process at work, which contrasted with the centripetal politics that predominated in the French context.[39] There was much less discussion in France precisely because there was much more common agreement on what the economic objectives of the industry should be. This did not mean that there were no differences of opinion - there were and often very bitter ones, as will emerge in the next chapter - but these differences revolved around how best the state's accepted role as protector of the whole industry should be developed. The result of the contrast was a marked difference in the way in which national interests, were conceived and pursued in the two countries. Whereas in Britain there was great stress on the importance of consultation and consent as the only way of uniting a cacophony of voices, in France the procedures were much more institutionalised, with the industry itself able to play a part in devising a national response under the supervision of the 'tutellary' state. These contrasting roles will be considered in the final part of this section.

3.1 Deep-sea and inshore interests

One of the major questions thrown up by the history of the fishing industry in recent years is why Britain accepted the terms of the CFP, as agreed in 1970. It seems remarkable that whatever the enthusiasm of the Heath government for joining the Community, it was possible to reach an agreement which was to pose so many problems. What is often forgotten is that the shape of the industry at the beginning of the 1970s was quite different from what it was ten years later. By 1980 it was the Scottish inshore ports that were at the centre of the industry, whereas in 1970 it was the deep-sea industry concentrated in Hull and Grimsby which predominated. The importance of this change lies not just in terms of the industry's structure but also in terms of the character of the message that the interests in the industry transmitted to government.

At the beginning of the 1970s the inshore and deep-sea sectors made no serious attempt to coordinate their positions in the face of the prospect of British membership of the EEC. The inshore industry was worried at the prospect of fishing 'up to the beaches' which the 'equal access' provision of the 1970 agreement appeared to promise. In October 1971, for example, they dropped in on the Brighton Conference of the Conservative Party from 20 or so boats "with anti-EEC slogans and Very pistols."[40] The owners of the deep-sea companies, for their part, saw little reason to join forces with their small-scale colleagues. The prospect of Norway joining the EEC made them far from unenthusiastic about a policy which prejudiced the chances of any member state creating exclusive limits. Norwegian waters offered a very attractive proposition to the men of the BTF and STF.[41]

They both had good reason for making their different perspectives known but there was no kind of pressure put upon them by the state to seek to reconcile those differences. The emphasis was on divergent economic interests being fully expressed rather than on mechanisms for fitting such divergences into a wider economic framework. The climate in France, protectionist in both an economic and a political sense, meant that it was possible to go some way towards transcending this division of interest in the industry. There was an early acknowledgement that 200 mile EEZs were soon to come[42] and the 1972 agreement with Canada reflected the state's determination to seek to adjust to that change whatever the preferences of the firms involved. Similarly, the CCPM was able to vary fuel aid as between deep-sea and inshore interests, even agreeing to exclude the largest vessels of 'la grande pêche', who could pass on the increase in price thanks to the fixed contracts with fish processors that were a central part of their operations.[43]

In Britain, continued resistance to the idea of 200 mile EEZs meant that when the final 'Cod War' blew up in 1975, the deep-sea industry was once again able to give full expression to its desire to maintain access to Icelandic fishing grounds. Despite the fact that the government was engaged in negotiations with the EEC over the extension of limits around Britain's own coasts, it used Royal Navy ships in a very costly defence of the right of trawlers to operate around Iceland. What is more, it was one that proved a disastrous failure. After Icelandic offers of 65,000t. and then 40,000t. a year were rejected, an agreement was reached in Oslo in June 1976 allowing access to the end of the year at a level equivalent to an annual catch of 30,000t. There was no further accord: British trawlers were suddenly debarred from waters that they had frequented since the 1880s.

As with the debate on the CFP, this remarkable chapter in fisheries history prompted a major inquest. Many felt that an opportunity had been missed and the chance of maintaining some distant-water operations around Iceland needlessly thrown away. A former ambassador to Iceland commented:

> "It is my belief that if we had made the best settlement possible ..., we would still (1978) be fishing in Icelandic waters today on a quota basis of about 40,000 tons."[44]

His own explanation for the conspicuous failure to reach such an agreement was set at a high level of generality. He suggested that the government's constant reiteration of the strength of her claim in international law was a reflection of a total lack of awareness of how the world was changing, with Britain's role in it undergoing remorseless decline.[45] Others were prepared to argue that it was the fault of individuals: Hattersley, Minister of State at the Foreign Office, responsible for the negotiations, and Austen Laing, Director-General of the BFF, were favourite scapegoats.[46]

However, an explanation can be found at an intermediate level. What was critical was the relationship between the state and this section of society. While for the British state it was important to be seen to respond to the calls of the industry, amplified in the parliamentary arena, the deep-sea owners were defending their economic interests as they saw fit. Almost no-one at the time contested what was being done, except to say that the government was not using its military muscle as effectively as it could. Certainly the state made no attempt to impose itself by arguing that it would be better to bow to political reality, make the best deal possible and seek other ways of redeploying the fleet. At the same time, the rest of the industry stood on the sidelines as it could see no good reason for becoming involved. That the situation was of this kind is not something for which any individual can be blamed, it was a necessary product of a political process where an instrumental view of interests plays such a predominant role (cf p 20 above).

The same centrifugal pressures can be detected in what occurred as the deep-sea industry progressively disintegrated. There was very little love lost between the old and new centres of power of the industry. The inshore men did not forget the disdain with which the big owners had treated them in the past and now argued that it was ridiculous to give subsidies to owners who were no longer going to send their ships to sea (cf pp 168-9). The company men, for their part, went out to cut their losses as best they could. Even within the deep-sea industry there was a feeling that what was happening reflected its inability to continue to protect its own interests and that its disappearance was an inevitable consequence of that inability. As Trawling Times, the paper of the BFF, put it in the final issue in May 1979:

> "In the tradition of this industry there are no crocodile tears ... We can no longer earn our keep. So we go."[47]

The state for its part, whether in the guise of a Conservative or Labour government, was not concerned to modify that tradition. If the deep-sea vessels were to disappear no British government, nor indeed anyone else in the industry, was going to prevent that disappearance out of a desire to maintain the existing shape or productive potential of the fishing industry.

3.2 Employer and employee interests

The apparent unity of the British industry was not only breached by the divide between the inshore and the deep-sea sectors. There was also a very bitter conflict between employer and employee interests, dominated by the TGWU's concern to obtain the decasualisation of the industry. Relations between the owners and the union were anything but cordial. Cairns, the National Fishing Officer of the TGWU argued that the employers were the worst he had ever encountered: "the last bastion of reaction" was his phrase.[48] The owners returned the compliment and went out of their way to snub the union. Before a meeting of the Sea Fisheries Training Council, set up by the Manpower Services Commission, the Grimsby owners let it be known that they would not attend because Cairns was going to be there.[49]

The result of this niggling between the two sides was that even under the difficult conditions of the late 1970s, it proved very hard to get any kind of common stance from this section of the industry. The stress was on ventilating grievances not seeking to resolve them. It was something of a minor triumph when the Humberside County Council, in a report on the state of the fishing industry in the region, managed to incorporate a paragraph arguing that the hiring of fishermen was a throwback to the

Victorian era, and that they needed to be given the same protection as other workers. While Cairns insisted upon it, the owners only agreed under protest.[50]

At a more general level, the two sides became deadlocked over the issue of the casual recruitment of labour. The TGWU got the three Hull MPs to ask MAFF to draw up plans on decasualisation right at the beginning of the Labour government in 1974 but little headway was made. The immediate response was that the issue was one for the industry to sort out and not the government.[51] Given the attitude of the owners, there was little to be done and, once the Conservatives returned to office in 1979, still less. Even within the union, however, the issue was one that did not command unanimity. There were those who feared that the introduction of a national basic wage would encourage trawler owners to concentrate their activities on Hull and Grimsby. Bernstone, the TGWU District Officer for North Shields, for example, argued that "the Transport and General Workers' Union policy, if it was applied absolutely to North Shields, would be the death knell."[52]

At the same time, there was no question of the decline of the deep-sea industry being accompanied by closer links between the TGWU and the inshore industry. Their differences of opinion were openly expressed and no attempt made to contain them. Further to an EEC proposal for improvements in social benefits, including the right to a pension at the age of 55 (as in France) and limits on excessive periods of continuous working, a union official went up to Peterhead to distribute a leaflet outlining the proposals and to debate the matter:

> "Most crew members and many skippers expressed an immediate interest in the pension proposal. A number of skippers also expressed an immediate interest in throwing me in the dock. Most skippers and indeed a number of crewmen felt that any regulation applying to working hours would be impracticable."[53]

This union involvement in the affairs of the inshore industry was resented by the SFF. Its Chief Executive replied that the union was misrepresenting its views and confusing two points:

> "The one which envisages the eligibility of a fisherman to a pension at age 55 and the other which represents an attempt on the part of the trade union movement in Europe to impose upon the shore-operated sector of the fishing industry a set of working conditions which are totally at variance with the whole concept of share fishing and co-adventureship."[54]

Whereas the SFF strongly favoured the early pension proposal, it could not possibly accept the idea of any restrictions on working hours. Such limits constituted a challenge to the whole ideology of the inshore sector, which had no desire to be fettered by the restraints applicable in other sectors of the economy (cf p 64 above).

Such a level of friction and such isolation of the unions was less marked in France. It is true that membership was not extensive and that rivalries between unions, notably the CFDT and CGT, undermined the common front of the employees but nevertheless the atmosphere was very different. In part, this was because the maritime sections of the unions were firmly incorporated within the framework of the 1945 institutions and saw them as an effective means of furthering their interests. Thus it was at the initiative of the chief CGT representative on the CCPM Bureau that a special committee was established to watch over the evolution of the number of ships in the fishing fleet, the Commission Nationale de la Flotte de Pêche (cf p145 above). This offered a channel within the existing framework for union anxieties about the industry as a whole to be expressed. Similarly, the CFDT was concerned to use its position inside FIOM to point to the overall distributive effects of the aid policy of the government. Despite its own low membership amongst the 'artisans', it argued that far too much money (7,239 FF per man) was going to the deep-sea industry, when it was compared with 215 FF per man being paid out to the inshore industry.[55]

However, tension was also less because the unions recognised that fishermen enjoyed a social regime more favourable than that of most people on land. This did not mean that there were no attempts to improve the position. In Etaples, paid holidays and compensation payments for the loss of earning were instituted for inshore fishermen and in some deep-sea ports agreement was reached with the owners to establish written records of hours.[56] However, these changes only served to underline the contrast with other EEC countries, including Britain. The difference between national industries became very clear when union representatives from several countries attended meetings in Brussels and pressed for the use of the Treaty (Article 117) to develop social policy in the fisheries sector.[57] The French representatives were amazed to find that the representatives of other industries, like the SFF, wanted to set minimum standards based on the lowest level of provision rather than harmonising upwards to the highest level.[58] From this comparison it is clear that the employer/employee clivage in France did not challenge the cohesion of the industry in France to anything like the extent that marked the British case.

3.3 Regional interests

It was suggested above (p 207)that the trips of the British fishermen to Brussels illustrated a united ability by the industry to develop new techniques to influence government. However, the number of people making the trip to the bar of the Charlemagne building in Brussels was also an indicator of the regional tension within the British industry. This tension continued throughout the period of this study, despite increased organisational unity. A good example of the difficulty could be seen in the activity of the SFF. The Federation spans an enormously large range of geographically dispersed fishing communities with considerable suspicion between them. Thus as many as six men went to EEC meetings from the Shetlands because they were determined to ensure that their own particular point of view did not become submerged in the overall stance of the Federation.[59] Indeed as the EEC settlement came to be formulated, the SFF was threatened by a substantial breakaway of component associations including the Shetlanders: the alliance that the fight against Brussels had forged proved fragile with the ending of that fight.[60]

Similarly, there was a running battle between the 'nomad' Scottish fishermen, who descended to the South West of England in search of mackerel and the small-scale local boat owners of Devon and Cornwall, who felt that their own grounds were being put at risk by the massive catches of the men from the North. And, it proved very difficult indeed to devise any way of dampening the conflict. It began as early as 1975 and was still rumbling on five years later, complicated still further by the presence of the big trawlers from Hull, whose owners saw in mackerel one way of keeping their boats in operation.[61] The issue illustrated very clearly that simply calling for a fairer EEC settlement generated no deeply-rooted consensus. As early as April 1976, the Editor of Fishing News commented that many in the industry would be well satisfied by the Commission's proposal for a 12 mile exclusive limit.[62] The owners in the South West were an example of a group for whom the calls for 200 or 100 mile limits were both more and less than they needed. They could expect to keep out foreign boats with a much smaller band but they could not be protected in such a way from fellow nationals. Such protection required something more than the winning of a lion's share of fish for Britain within the EEC.

Such divergences in the industry were amplified as a result of the way complaints were ventilated: the parliamentary arena not only gave greater salience to the issue as a whole, it also accentuated the competition amongst fishermen. An example can be seen in the way in which the Hull MPs protected their own port's position in relation to Grimsby.[63] The decline in the number of ships fishing from Hull had the effect of pushing up the cost of unloading as the use of the dock was spread among fewer operators. By 1980, the cost had risen to £51.92 per tonne, as compared with £10 per tonne in Grimsby, where the presence of an inshore fleet had served to keep expenses down. However, Prescott, in particular, made full use of this comparison on the radio and was successful in getting the British Transport Docks Board to bring the cost down to £22.50 per tonne for British trawlers and £12.63 for Icelandic ones. The intention was to encourage foreign ships with their very big catches to be more willing to come over from Grimsby to Hull. It was a classic example of the decentralised character of the policy process in Britain, very flexible but with little conception of harmony among competing private economic interests.

It is true that the French fishing industry has never been as concentrated regionally as the British one. In France there is a greater evenness reflected both in the lack of a single dominant port or area and the relatively small size of the deep-sea companies, when compared with their British counterparts. Nevertheless, this lesser degree of concentration does not mean that there was no conflict between regions, the most conspicuous being between the North, centred on Boulogne, and Brittany. In 1971, for example, the Breton owners clashed with the men from the North, because the former felt that the latter were pushing too hard in the debate over the shareout of money for modernising the fleet.[64] As a result the Bretons established their own association, though they did not carry through their threat to leave the UAP altogether. Subsequently, there was a significant degree of mistrust because Boulogne was the main source of the imports that came into France, while Brittany was the area that suffered most from the drop in prices caused by those same imports.

However, these conflicts did not spread or increase in visibility during the period of this study. The reason was that individual regions were not free to protect themselves at the expense of others, by changing their pattern of behaviour in a major way. When there were conflicts between particular ports as between La Turballe and Quiberon in Brittany, the fishermen did write to their local deputy, Bonnet, to give their case more force but the solution still came through traditional channels. The local maritime administrator brought the representatives of the two local committees together and got them to agree on new rules, without the issue gaining any further ventilation.[65] On a larger scale, it has already been shown how disagreements between areas were contained within the central state-sponsored institutions. Inside FIOM the owners of the North were defeated in their attempts to limit the money going to the SNPL from Bordeaux (cf p187above). The vote against them did not eliminate the conflict but equally there was no question of it being challenged within the parliamentary arena, as it would surely have been in the British context.Nor was this centripetal process one that the industry as a whole seriously resented: it was very ambivalent about the notion of more local autonomy within the industry. When Le Pensec presented the CCPM with his proposals for change in the context of the government's general policy of decentralisation, the response of the profession was anything but favourable.[66] The Ministry in Paris was to retain control of most decisions - ENIM, training, employment and security issues as well as disciplinary and general administrative matters. However, there was considerable disquiet about the facts that aid submissions for building boats would be devolved to the regions, and that the administrative structure of 'Affaires Maritimes' would be placed directly under the prefects, renamed 'Commissioners of the Republic'. The industry feared that it would lose its direct links with the Ministry in Paris. To a man, the Bureau of the CCPM - owners, unions, inshore, deep-sea - gave a fulsome tribute to the work of the maritime administrators and pressed for the status quo to be retained.

Once again the example indicates the importance of deep-rooted societal attitudes towards the state. In the fishing industry these attitudes provoked resistance to modest change whilst at the same time underlining the cohesion of the industry. There was a reluctance to accept that regions should have more of a say in their own affairs. This was in marked contrast with the British mediatory tradition, where differences between regions could be freely expressed and amplified in the open parliamentary arena.

3.4 National interests

The difference between the two traditions is also well illustrated in the way in which broad national interests were conceived and pursued. Contrasting economic stances in the two states combined with distinctive political styles to generate approaches to the identification of national interests which bore very little relation to one another. Here we will illustrate the contrast by looking first at the way in which the British government moved towards a settlement of the fisheries issue within the EEC and then at how the French responded to the arrest of a number of Breton trawlers off the South Wales coast.

When Walker presented the EEC agreement to the Commons in January 1983, he was able to describe it as "superb", without provoking a major outburst of indignation.[67] The majority of the those who spoke were Conservatives and they congratulated the Minister on his success. Yet even two Labour members, Johnson of Hull West and Brown of Glasgow Provan, were appreciative of his efforts in getting a deal which was, in their view, as good as one could expect in the circumstances. When the Labour front bench spokesman, Buchan, pulled out his copy of Fishing News with its claim that 98% of fishermen did not believe that the new CFP provided proper controls, Walker was able to put him down very easily by saying that 69,880 of the paper's 70,000 readers had not voted in the poll!

The change of atmosphere is a remarkable one when one considers the level of unanimity that a belief in the rightness of Britain's case had engendered from 1976 onwards. Though demands for exclusive limits were progressively reduced from 200 to 100 to 50 to 12 miles, the spirit that accompanied these retreats was one of grudging acceptance of political reality, rather than a belief that they were justified in terms of fairness. Comparisons with Norway which had rejected membership of the Community and with Iceland which had rejected the British deep-sea industry in 1976. were ones that implicitly or explicitly influenced thinking throughout the period.[68] After all, both of them had been free to declare their own EEZ without reference to anyone else.

There was a similar level of unanimity on the issue of what percentage of the fish in EEC waters British vessels should be allowed to catch. An SFF newsletter put the point as follows:

> "As the nation in whose waters the bulk of the fish swim (over 60%) simple justice requires that in the eventual CFP settlement, we be accorded a share of the resources to which we contribute which fairly reflects that contribution."[69]

It was a position which no-one was eager to contest. Even the European Communities Committee of the House of Lords, which was of the opinion that the difficulties of the industry should be attributed to the loss of Icelandic waters rather than membership of the EEC, called for 45% of the fish in EEC waters to be granted to Britain rather than the 31% that was on offer at the time in

1980. The Daily Telegraph, no opponent of the EEC either, backed the Committee's call objecting to any doctrine of common resources, which might challenge a state's sovereign rights

> "... a country's fish stocks may be regarded as its property whatever it makes of them; and there is no clear reason why membership of the EEC should entail international socialism."[70]

Such appeals to national sovereignty, a focus of unity for all shades of political opinion, swamped alternative definitions of the problem and of the possibilities that a CFP might offer. There was limited discussion of the use of the Community price structure to protect fishermen, despite the fact that before membership "British fishermen for years (had) wanted a minimum price scheme which would come to their rescue when the bottom falls out of the market."[71] There was little awareness of the benefits of developing the terminology of the Community, such as 'coastal preference', so as to win arguments in the Council of Ministers.[72] There was minimal recognition of the significance of fish as a migratory resource, where a change in fishing patterns could alter catch distribution very markedly (cf p 87 above). And there was almost no acknowledgement of the costs that would be imposed on fishing communities such as those in Brittany if the waters around Britain were to be closed to them. The politics of the virility symbol prevailed.

In the EEC arena no British government sought to change the definition of the problem, rather they all made the principle of consultation and consent central to the way negotiations were pursued. When Silkin was in office under the Labour government, he could argue that a 12 mile exclusive zone, 'dominant preference' between 12 and 50 miles and national conservation measures up to 200 miles was the least that the industry could be expected to accept and that anything less would be a sell-out. However, Walker was able to use the attitudes of the industry in the same way to deflect domestic criticism and to move towards agreement. He did this by maintaining very close consultation with the leaders of the industry and by modifying the criteria for an acceptable settlement.

The importance of the widest consultation with the fishing industry was something which the Conservative government after 1979 was very careful to stress. As Buchanan-Smith, Walker's Minister of State, put it:

> "We have representatives of the industry in attendance at Fisheries Councils and we consult them during the negotiations. I appreciate the help and support given by the industry in relation to this matter. This is something which we shall certainly continue and we shall work very closely with it."[73]

Subsequently Walker proved remarkably successful in developing a close relationship with the nine main industry representatives.[74] He revelled in their presence in Brussels and got them to accept concessions bit by bit, persuading them to move ever closer to the government position and get their own members to do likewise. By the end, it was difficult for them to say that they had been sold-out, as they had been intimately involved in what had been going on.

At the same time, Walker recognised the value of consent in the wider political area. In the 1979 elections, the Conservatives had been just as tough as Labour, if slightly less specific, in the conditions that they had laid down for a satisfactory settlement of the CFP.[75] However, what Walker did was to add the consent of the industry as a further condition for a settlement. Thus though the agreement was not enormously different from what would have been available in 1976, at least by comparison with the earlier demands of those inside and outside the industry, it was very difficult to attack Walker. He was able to claim - and rightly - that he had won the support of the leaders of the industry for that agreement.[76] The result was the identification of a national interest through the use of informal persuasion and cajoling.

The contrast between the state's role in French and British experience is well illustrated by what happened in France in the autumn of 1979 following the arrest and fining of two Breton trawlers off the South Wales coast for using nets with illegal mesh sizes. Though the issue was not as major as the question of the overall shape of the CFP, it was similar in that resolution of the problem depended on factors outside the national arena and the degree of domestic unanimity was every bit as great as that which prevailed in Britain.

There had been considerable worry in the French fishing industry as to the effects of the introduction of EEZs, because of the possibility of unilateral British action. In 1975 one Lorient owner explained the fact that he put his map of British waters in a cupboard by saying:

> "I hide this away for when the fishermen see it they go berserk with panic ... We desparately hope that Britain will agree to the Nine sharing their 200 mile zones on a Commmunity basis."[77]

On a visit to Brittany in February 1977, Giscard d'Estaing responded to such worries by underlining the government's commitment to ensuring that fishing would continue around Britain:

> "I have asked the Ministers of Transport and Foreign Affairs to be intransigent in the defence of French fishing rights. The fishermen are not to worry. Their traditional rights to fish will be acknowledged and protected."[78]

When in July 1979, the British government introduced a 70 mm mesh size for nets used to catch Norway lobster ('langoustine'), there was an immediate reaction in France whose extent underlined the level of unanimity on the issue of 'historic rights'. All, including the newspaper of the PCF, were forthright in their attack on Britain: the principle of conservation was being used, it was said, to hide a narrow nationalism.[79] However, there was more than just newspaper comment. On July 19, the Bureau of the CCPM agreed that it would pay any fines imposed on French trawlers by British courts.[80] The latter reaction was particularly significant in that it underlined the way in which the state could subscribe to a decision of the whole industry to protect one section even before the effects of the British action had been felt or any effort had been made to reverse the decision through diplomatic channels. In Britain, over the EEC settlement, the reverse was true: diplomatic channels became the focus of attention with no semblance of domestic unity as how best to develop the different parts of the fishing industry. Such development would have to await a settlement whatever its anticipated economic impact.

The first overt diplomatic move in this case did not come until the middle of September after the first two arrests. Le Theule, the Minister of Transport, wrote to Walker charging Britain with acting against Treaty obligations and hence illegally.[81] Walker retorted that the mesh size used by the boats in question would have been illegal even before the July measures and would remain so under the proposed EEC regulations, due to come into force at the beginning of 1982.[82]

However, this diplomatic activity became increasingly irrelevant to domestic French reaction. On October 1, the CCPM decided to use the industry's funds to support arrested boats to the level of 8000 FF per day to cover their loss of earnings. There was considerable dissatisfaction that the state itself in the shape of the Finance Ministry was not prepared to provide the money out of public funds. As the CCPM put it, the state should:

> "protect its members and repay them the losses they incur when those concerned are not breaking the relevant French regulations, which the Commission in Brussels has also approved."[83]

Indeed it was even felt that the British state should be responsible for reimbursing the French fishermen if the arrests were declared illegal by the ECJ in Luxembourg. There was no chance of this but in the end the French state did contribute 53,000 FF towards the cost of the six boats in all that were arrested and charged, the profession itself finding 157,000 FF.[84] However, it was not just that the French state was prepared to offer economic protection to fishermen as part of its definition of the overall French interest. There is a further contrast with the British experience. French fishermen themselves set only limited value on the existing consultative framework and informal consultation on the Walker model was not available as an alternative. There was a clear move towards action outside the conventional channels. Initial strikes by the fishermen of the 'pays bigouden' and occupation of a townhall and the offices of the maritime administrator in Le Guilvinec was followed by more robust action. The car ferry at Roscoff was occupied and the fish cargo of the lorries on board sprayed with oil or thrown into the harbour, causing 200,000 FF of damage.[85] It was only a short lived affair - by the beginning of October, the fishermen were back at sea - but it pointed to the possibility that French fishermen could seek to make their views known through "a much wider set of channels", though very different ones from those pursued in Britain.

4. Conclusion

The conclusions of this chapter are threefold. Firstly, it has suggested that the importance of the British paliamentary arena meant that 'wider channels' were available for the leaders of the British industry to exploit. The scope for action of the conventional, consultative structures in France was more limited. At the same time, that scope for action was strongly influenced by the industry's attitudes towards the state and the state's own role. In the French context, the shared stress on an aggregate, public interest restrained wider discussion and the extensive ventilation of grievances; in Britain, the full articulation of separate private interests in the industry was encouraged and the state was not concerned to prevent it. Hence the validity of the distinction offered in the second of our initial assumptions helps to explain why the first assumption is not confirmed in the French case. (cf p191)

The second conclusion is that the material here does not seem to confirm the general argument that interest groups are necessarily supplanting traditional forms of representation, such as political parties.[86] The pattern was very different in the two countries discussed here. In Britain, parties played an important role in buttressing the consultative ethic to which all subscribed, whereas in France their role tends to confirm the argument that:

> "... the transfer of power from legislature to the executive with the formation of the Fifth Republic has resulted in less pressure by interest groups on political parties and a concentration at the executive/bureaucratic levels where policy is decided."[87]

Finally, the chapter has pointed to the importance of different mechanisms for reaching decisions, each with its own weaknesses. The British stress on compromise between conflicting economic interests opens it to the accusation of a lack of decisiveness. As Hayward has put it:

> "... in Britain the presumption that group consent is an indispensable prelude to action has generally resulted in timid and irresolute decision-making."[88]

The French enthusiasm for common goals has not encouraged timidity but it has meant a less developed sense of the need to win over groups through persuasion and discussion. As we shall see in the next chapter, these two approaches are subject to different pressures when an interest goes outside the conventional channels for exercising influence. In France centripetal pressures created the potential for a very large explosion of discontent, in Britain centrifugal forces helped to diffuse disaffection and to make a virtue of 'muddling through'.

7 The direct-action perspective

1. Introduction

This chapter takes up the issue raised at the end of the previous chapter, namely the abandonment of the conventional channnels of consultation and the development of techniques of direct action as a way for the fishing industry to influence government. It will examine the nature of that action in the light of what we might expect to have occurred, as outlined at the beginning of the study (pp 30-1 above). It was suggested, first, that the result of major change in the issue area would be for the industry to be pushed into "forms of direct action in as far as it perceived that its demands for assistance made no headway through the conventional channels." We will find that actual or threatened direct action did become an important element in government-industry relations for the period of this study, as it had not been before. We will also find that the threats were carried out more often and more vigorously in France. Hence the evidence supports our second suggestion that "the importance of the search for consensus in British policy-making and the 'limited authoritarianism' implicit in the French activist policy style would make ... direct action more common in France than it was in Britain."

The chapter is divided into three sections. The first will consider the frequency and form of direct action concentrating on 1975 and 1980/81, when the dissatisfaction of the industry in both countries was at its peak. The next two sections will look in turn at the industry and the state during these periods of unrest. It will be argued that the behaviour of the one needs to be understood in terms of the behaviour of the other. The French fishermen's frustration with the traditional state-sponsored mechanisms of representation found expression in protest that spread wildly in a spontaneous and uncontrolled way and the state's response was marked by a powerful desire to reestablish unity and order whatever the cost. The British state's concern was rather one of winning over the fishermen with limited concessions, directed at a representative leadership. That leadership channelled the discontent of its members in a very deliberate and organised way.

In conclusion, it will be suggested that the balance between centrifugal and centripetal forces in the two countries discussed in the previous chapter was reversed in the context of direct action. In France, as a rule, the power of centripetal forces was far greater than that of centrifugal forces. But as this chapter shows, their position could be drastically reversed and the tensions within the 'one and indivisible Republic' laid bare. In Britain, the state was disinclined to aggregate interests and tended to allow full play to centrifugal forces. At the same time, when direct action was threatened or carried out, the representatives of the industry were able to restrain their members from going too far and government representatives were willing to seek some form of accomodation with the protestors.

2. Direct action: its frequency and form

The history of the fishing industry is punctuated by strong conflicts of interest, which were by no means always resolved through conventional channels. In 1896 there were riots at Newlyn in Cornwall because the local fishermen were furious with the ungodly men of Lowestoft for working on Sunday. So severe were the fights between the two groups that it took the arrival of 350 soldiers to quell them.[1] More recently, there were important strikes in the deep-sea industry, which hit both Britain and France. Stoppages at Grimsby and Hull in 1958 and 1961 were followed by ones at Concarneau and Boulogne in 1967 and 1968. Within the French inshore industry, the first part of the 1970s witnessed a number of small-scale blockades of individual ports, prompted by specific local grievances.[2]

However, during the period of this study the expression of discontent took a different shape. First of all, it assumed a national rather than a local significance: sectional complaints were amplified and won support amongst a much broader range of fishermen. Secondly, this new unity provoked serious conflict with the state. The authorities were obliged to find a response to tactics, such as the use of blockades, which were designed to prevent normal maritime movements through illegal means. Thirdly, the justification of such direct action was set in terms of the relative position of fishermen in relation to others: militancy was seen as the one way of getting one's voice heard in a world where the government would only concede to the demands of the strong.

In more general terms, direct action within the fishing industry formed part of a secular trend within all European democracies. More and more people have wanted to have their say but not everybody has been able to get a hearing. The problem was particularly acute in a period of resource 'squeeze' when the possibilities for governments to buy off discontent were reduced. It is significant that the use or threat of direct action in the fishing industry was most prominent in 1975 and 1980/1, when increases in fuel prices were making their presence felt throughout the economies of Western Europe and when both Britain and France had governments committed to control of public expenditure. As Richardson has pointed out:

> "... the policy process in the period of steady expansion in public expenditure was not unlike the Dodo's race in Alice in Wonderland in which everyone won and everyone got prizes. Not surprisingly a race in which there are winners and losers is more difficult to manage."[3]

The fishermen did not want to be losers but their governments were not free to continue to give everyone prizes. How this potential conflict was resolved we will consider by looking chronologically at the examples of the use of or threat of direct action during the period of this study.

2.1 France - 1975

As we have seen (p164above) the French government responded to the increase in the cost of petrol in 1973 by introducing a fuel subsidy which amounted to 20MF in 1974 and was set at 12MF in January 1975 for the coming year.[4] At the same time, there was a recognition that one of the indirect effects of the increase in petrol prices was an increase in imports and a consequent drop in price levels obtained at auctions (cf p 43 above). As a result, Cavaillé, the Secretary of State for Transport, called upon the EEC Commission to improve its mechanisms for temporarily restricting imports.[5]

However, what followed underlined that for many in the industry these measures were not enough. The 'artisans' of Etaples, near Boulogne, had been obliged to stay in port for 8 weeks because of bad weather and they were furious when they found that competition from imports helped to push their first catches of the year in February 1975 to very low price levels. What is more, the state's treatment of them appeared unfavourable when contrasted with its attitude towards the farmers. On Saturday February 15, Chirac, the Prime Minister, announced that he could not contemplate a drop in the earnings of farmers for two years in succession and promised important subsidies involving practically a 13th month of salary.[6] The following evening the trouble began when the 300 inshore fishermen of Etaples were persuaded by their leader, Bigot, to travel down overnight by coach to Paris to vent their grievances. On arrival in the capital the protestors occupied the entrance of the Ministry, the SGMM, in the Place Fontenoy. The Minister refused a request to see them and on the arrival of the police the men were ejected at the cost of some minor injuries and a few broken windows.

Their subsequent departure from Paris was to prove the beginning rather than the end of the affair as sympathy action spread throughout France. On the Tuesday, the channel ports of Dunkirk and Boulogne were blockaded by inshore boats and in Brittany, deep-sea crews from Douarnenez, Lorient and Concarneau came out on an 8 day strike in support of the men from the North, calling for an increased fuel subsidy and the closing of the frontiers to imports. By the Wednesday, nearly every port from Dunkirk to St.Jean de Luz on the Spanish frontier was blockaded and fishermen from Calais had carried the action further by intercepting fish lorries from Belgium and Holland on their way to Paris and dumping their contents on the motorway.

The dramatic spread of the dispute encouraged the government to sit down with the fishermen. On the Thursday, Cavaillé did accept to see a delegation from Brittany and the North. The meeting was a tough one but after five hours of negotiations, the government made important concessions, agreeing to further financial help: the fuel subsidy for 1975 was increased so as to be worth about 4 to 5 centimes per litre; an extra 11MF was set aside as social assistance for the inshore fleet to compensate for its drop in purchasing power; 20MF was provided to be made available for the financing of withdrawals from the market by the producer organisations, set up under EEC regulations; and limits on imports from non-EEC states were to be introduced on the basis of the safeguard clauses in the Treaty of Rome.

Despite this acknowledgement of the importance of protecting the industry, the unrest did not end straight away. On the Mediterranean coast a blockade was maintained at Sète until the end of March, where the men had a particular complaint about the levels of pollution. In Brittany the deep-sea crews took the opportunity of pressing for better wages and working conditions, extending their tactics beyond those of the traditional strike. On Monday, March 3, crews from Lorient copied the men of Etaples and descended on Paris to demonstrate and to dump 40 tonnnes of fish in the smart 16th district. Even then the protests did not end straight away. It was not until March 21 that the longest-running strike, at Concarneau, ended and the men returned to sea.

2.2 Britain - 1975 and 1980/1

The grievances of the British industry in 1975 were not unlike those of its counterpart in France. The cost of fuel had jumped dramatically since 1973 and imports were having a major effect on prices at the auctions. Moreover, as in France, the government had made an initial response to these difficulties. On February 27, Bishop, Minister of State at MAFF, announced in a written answer that a temporary operating subsidy of six and a quarter million pounds would be introduced for the period from January 1 to June 30.

However, supporters of the industry were not slow to point to the extra help that the French government had agreed to, particularly the restrictions on imports. As Buchanan-Smith put it in the Commons: "how do we explain to our fishermen if the

French can do it, why the British Government can not do it as well?"[7] In the same debate Gray of the SNP felt that the relative success of the French fishermen would tempt, even justify, their British counterparts to act similarly unless the government showed more concern for their case:

> "It would be a great pity if our fishermen had to resort to the extreme measures which their French neighbours took to gain some sort of Government recognition."[8]

Self-imposed restraints upon their tactics were not only weakened by what they saw happening in France. Fishermen also felt that others in Britain were getting their way by strong-arm tactics. Rather than looking at farmers, as their French counterparts did, they compared their situation with that of workers in the secondary sector where:

> "... the Government continued to pour millions into unprofitable industries; whenever a union seemed strong enough, it appeared that the Government would concede whatever was demanded."[9]

This combination of particular and general grievances was powerful enough to prompt direct action, some four weeks after the equivalent action in France. On Tuesday March 18, the newly founded Humberside Share Fishermen's Association decided to blockade Grimsby and Immingham and these blockades were put into effect the following day. Immediately the Tor-line ferry 'Anglo', with her 500 passengers from Gothenberg was prevented from docking. Despite informal contacts with the government, the blockade was maintained and by the Saturday, 65 boats and 260 men were involved.

The following day the North Shields fishermen decided to join in and that same afternoon they blocked the Tyne estuary. Up in Scotland inshore boats from Wick to Peterhead stayed in port on the Monday (March 24), while further blockades were established on that day at Blyth, Sunderland, Hartlepool, Whitby and Scarborough. In the Commons, Peart, the Minister of Agriculture, indicated that he had invited representatives of the fishermen to meet him on the following day. The day after the meeting, on the Wednesday, Peart made a further statement in the House but its rather general contents failed to calm the fishermen. An Action Committee was established at a Thursday meeting in Aberdeen, with representatives from all the main Scottish inshore ports plus North Shields and Grimsby. The Committee decided on a full-scale blockade to start at midnight on Sunday, March 30.

Meetings were held in all the fishing ports concerned to decide on tactics and careful preparations were made to ensure that the action had maximum effect. Despite a poor weather forecast on the Sunday evening, the Monday morning duly saw vessels converging from all sides on the selected targets. The boats of Peterhead and Fraserburgh, for example, made their way to Aberdeen and within one and a quarter hours of their arrival had completely blocked the harbour. The pattern was the same throughout Scotland, with 860 vessels involved at 18 separate locations from Granton in the East to Ayr in the West via Lerwick in the Shetlands. In England, too, there was support, particularly on the Tyne, at Hull and other ports as far apart as Newhaven and Whitehaven.

In general, there was little trouble though on the first day, on the Tyne, a 7,800 ton collier, the 'Chevington', broke the blockade, by rammming a small 56 foot inshore vessel, and on the Tuesday, a small coaster was brought in to port by a pilot, whose cutter was subsequently chased by fishermen in angry mood. Already on the second day, there were informal contacts with civil servants from DAFS, which led to a meeting in Aberdeen with Brown, Under-Secretary of State for Scotland on Wednesday, April 2.

This meeting produced a Press statement from the Minister with an appeal for the end of the blockade. There were no firm commitments from the government but there was a general feeling that Brown's promises should be given a chance. The blockade could, after all, be restarted. After a long debate it was agreed to inform the ships involved that they should return the following morning to their home ports. The ships duly dispersed with one final gesture of defiance as the men on the Tyne threw bags of paint and oil at the 'Chevington'.

In fact, no controls on imports were introduced and no changes were made in the terms of the subsidy provided to the industry. The one concrete act that followed the blockade was that the Norwegians were persuaded by the government to restrict their landings in Britain. But this did not alter the general impression that very little had been gained except public attention.[10]

These relatively meagre results did not provoke the wider use of blockades; rather protest action continued to follow more conventional channels. June 14, 1977 saw fishermen from all over Britain descend on London in their boats and succeed in stopping traffic from using London Bridge. Then May 30 1980 was declared a Day of Action by the English fishermen of the NFFO. Yet it turned out to be little different from the traditional lobbying of Parliament with a delegation going to Downing Street and fish being sold in London streets at knockdown prices to delighted housewives. Even when the effect of imports was to make trips to sea unprofitable, the response did not extend beyond ships tying up and refusing to set sail. This happened at Peterhead in August 1980 when prices fell to £2 per box and was repeated in February 1981. When the 250 boats at Peterhead were joined by nearly all the rest of the British fleet.[11] Even the fishermen of the South West, no friends of the Scots, joined in. However, on both these occasions, the prospect of aid from the government defused the situation and there were no more serious attempts to challenge the power of the state by such actions as blockades.

2.3 France - 1980

In France, by contrast, the period after 1975 did see the industry resort to direct action more regularly. Between February 1975 and the end of 1979, 13 cases of unrest were reported, culminating in the incidents provoked by the British arrest of Breton trawlers off the South Wales coast in September 1979.[12] However, even this, the most serious clash with the authorities since 1975, did not provoke a chain reaction in other fishing ports. It remained the expression of a sectional grievance, even though the CCPM acknowledged its wider importance by using money from the industry as a whole to reimburse the owners of the trawlers concerned. By contrast, what happened in August 1980 saw a whole variety of separate discontents coalesce into a nationwide protest, which exceeded in scale, duration and ferocity anything that was seen in either country during the period of this study. Its impact was such that one participant went so far as to call it the 'May 1968' of the fishing industry.[13]

The origins of the protest can be traced to the plan for the deep-sea industry that Le Theule unveiled after the meeting of the French Council of Ministers on April 2, 1980. This plan called for the modernisation of the fleet as a way of cutting the trade deficit in fish, but this uncontroversial end was to be pursued by very controversial means. The 30MF set aside for 'la pêche industrielle' was made conditional on costs being cut by the trawler owners. This meant a reduction either in the wages of the crews or in manning levels, both of which the owners set about negotiating. At Boulogne they tried to persuade the unions to accept a revision of the agreement, 'convention collective', which had been negotiated in 1975. The owners wanted the size of crews on large trawlers to come down from 22 to 18 and a 10% reduction in crews' wages to offset rising fuel costs. The negotiations proved fruitless and on July 22 the owners decided to keep their boats in port, effectively locking the men out from work. Some two weeks later on August 6 a mass meeting was held to discuss a revised proposal that the crews be cut from 22 to 20, provided there were no redundancies. The men almost unanimously rejected the idea and then made their way to the car ferry terminal, preventing sailings to Britain for the rest of the day.

This initial action served to spark off protest elsewhere, even though the complaints were different ones. At nearby Etaples the merchants refused to buy everything that had been landed and in Normandy, at Port-en-Bessin, a similar dispute with the merchants was compounded by dissatisfaction over the failure to appoint a new official to pay out sickness benefits.[14] The inshore men at both places came out in support of the men from Boulogne on August 11, deciding not to go out to sea. Two days later the dispute was intensified when the Norman fishermen blockaded the ports of Le Havre and Caen, a tactic which had spread by the weekend of August 16-17 to Calais, Boulogne, Dunkirk and Cherbourg. This tactic produced the first dramatic confrontation of the dispute when on Sunday, August 17, the Townsend Thoresen ferry, 'Free Enterprise', broke the blockade at Cherbourg to the delight of stranded tourists and xenophobic elements in the British press.[15]

The question then was whether the dispute would extend any further. The government was organising regional conciliation meetings and the CCPM was due to meet on Tuesday, August 20, to enable all parties to the conflict to meet around the table. At the same time, there was considerable doubt as to whether Brittany, the supplier of nearly half of France's fish, would join in. This was by no means certain because the Breton inshore fishermen remembered that their protests the previous year had not prompted sympathy action elsewhere. They also pointed out that the summer was their major fishing period, whereas in Boulogne August is a quiet time.[16]

With some hesitation, the two main Breton ports, Concarneau and Lorient, did vote to join the strike and they were followed by the inshore centres of the 'pays bigouden'. However, the major incident of the week occurred on the Mediterranean coast, where the oil port of Fos, near Marseilles, was blockaded. To the commercial losses incurred in a port such as Le Havre, estimated at half a million pounds per day, was added a threat to the nation's petrol supplies. Barre, the Prime Minister, appeared on television to make it clear that the government would not tolerate this challenge to freedom of navigation and the navy duly broke up the blockade with powerful tugs and watercannon.

In the course of the dispute the various separate grievances had crystallized around the single issue of an increase in the level of the fuel subsidy. Such an increase would help the deep-sea owners of Boulogne as much as the inshore vessels of ports like Port-en-Bessin. A whole series of national and regional meetings over the weekend of August 23-24 seemed to offer the prospect of going some way towards meeting this common complaint. Indeed such a possibility encouraged the Norman fishermen to lift their blockade of Le Havre for 24 hours but it emerged in the discussions that the issue of fuel aid was not open for negotiation. So on the Monday, Normandy became the centre of renewed protest: the fishermen went to blockade Antifer, near Le Havre, France's second oil port, while the blockades at Fos, Dieppe, Boulogne and Dunkirk were restored.

The government showed no sign of making concessions. On August 26, the navy was again used to break the blockades of Antifer and Fos. On the same day, Le Theule announced the availability of new funds for a wide range of problems, including loans on vessel construction and price support, but made no reference to the chief demand of the industry for an increased fuel subsidy. The following day the Council of Ministers met but the fisheries question was not even on on the agenda. The only reference to the uproar on the coast came under the heading 'miscellaneous items' with the government:

> "confirming its instruction to keep the major French commercial ports open, their activity not being concerned by the problems of the fishing industry."[17]

Such clear unwillingness to agree to the fishermen's demands was followed by a gradual weakening of the action. At Fos the fishermen decided not to renew their action for fear of endangering lives and by the end of the week the only ports at a standstill outside the North were Concarneau and Lorient in Brittany and Bastia in Corsica. The inshore vessels of the 'pays bigouden' voted to return to sea on Saturday, August 30.

By the beginning of September, only the deep-sea men of Boulogne and the inshore boats of Normandy were still holding out but their scope for action was rapidly diminishing. At Boulogne, the fishermen found themselves in conflict with lorry drivers bringing fish to and from the processing factories. They were obliged to restrict their blockade to the port area alone. The men from Port-en-Bessin gave up any idea of a further blockade of Antifer - their clash with the navy had caused them between 2 and 3MF of damage - and carried their protest on September 2 to Paris. However, the combination of a delegation to the Transport Ministry and a successful attempt to block traffic near the Eiffel Tower failed to win any concessions. The next day a vote was taken and it was agreed by the 'petits patrons' to go back to work despite the opposition of the crews. Similar opposition was expressed by the men in Boulogne on Friday, September 5, when they voted by 276 to 193 against a proposed agreement guaranteeing manning levels for the time being, provided the unions accepted a revision of the 1975 'convention collective' in the course of the following month.

During the following week the eighth of the conflict, the decision of the Lorient men to return to sea on September 9 left Boulogne completely isolated. The men faced ever greater financial problems, as banal but necessary as paying for their children's books and pens on their return to school. They scraped money together at football matches and fairgrounds but the situation was hopeless. Eventually, on September 17, another vote was taken and this time a majority supported a return to sea on the basis of a temporary agreement which put off everything until a new accord was reached.

In the deep-sea industry the action over the previous weeks had brought a solution to its particular problem no nearer. Indeed the issue of manning levels was to remain unresolved for the rest of the period of this study. As for the grievance that the companies shared with the inshore industry about the need for a better fuel subsidy, there was no action on the part of the government to match that of 1975. It was universally agreed that the scale of direct action in France in 1980 was in inverse proportion to its success. The papers variously spoke of "total failure", "almost no gains" and "empty nets".[18] Le Monde argued that success was the government's: it had gambled on being able to play on the divisions in the ranks of the fishermen and it had won.[19]

3. Direct action: the industry's role

The fact that the 1980 action in France had affected many British holiday makers, who were unable to get back to England without a lengthy detour via Belgium, encouraged comparison between the two countries. A certain C.J. Beaumont, who wrote to the Daily Telegraph on August 22, was clear about the difference:

> "The British worker's attitudes to the law is one of respect either natural or enforced. This attitude or approach to the law and its enforcement does not seem to prevail in France.
>
> Be you peasant, King or fisherman, you are under the law: this is the only way a civilised country can function."

It is not the present author's intention to comment on the level of civilisation in France but it will be suggested here that the incidents just discussed do need to be explained in terms of the relationship between the state and society, of which the law and attitudes towards it form a part. It will be argued that the direct action of the fishermen illustrates the different ways in which a state-led society and a society-led state process conflict. In this section, the stress will be on how the grievances of the industry were represented to the state. In the following section, attention will turn to the response of the state. The assumption throughout is that the two processes were interlinked, the state sharing responsibility for the way grievances were represented and the character of the industry influencing the state's behaviour. The section will not deny that certain conjunctural variables, notably the prospect of the 1981 Presidential elections in France had an independent influence.

But the prospect of elections did not determine the basic structure of relations rather it exacerbated certain tensions within that structure.

3.1 The mode of organisation

One of the remarkable features of the direct action that was taken in France was the spontaneous way in which it spread from port to port. It was most obvious in 1980 when there were simultaneous blockades of ports as far apart as Fos and Antifer, but the same kind of development could be seen in 1975 as the grievances of the men of Etaples provided the occasion for others, like the deep-sea crews of Lorient to make their own discontents known. There was no central direction of these complaints but the sum of individual port decisions generated an overall picture of an entire industry up in arms at the way in which it was being treated by the state.

This decentralised structure of protest is in marked contrast to the situation in Britain where the major blockade of 1975 was under strong central control by the representative leaders of the industry. The initial blockades on Humberside and at North Shields were organised by the local associations through separate decisions, but the moves to the major blockade which followed Peart's statement to the House on Wednesday, March 6, were meticulously planned by the group of 19 men, known as the Action Committee. This group itself was set up at a meeting where 70 representatives of the industry met. They gave the Committee absolute powers to take decisions binding on all the ports involved. The Committee in turn elected a Chairman, Willie Hay of

Buckie, and then decided on the way in which the dispute would be pursued.

These arrangements defined very clearly the extent of the action that followed. There was no question, first of all, of a link-up between the inshore and deep-sea industries. A representative of the Scottish owners, the STF, stated his opposition to a blockade at the initial meeting and Hay maintained that the problems of the deep-sea crews themselves were different, in that they could not choose whether or not they went out to sea.[20] As for the unions, the Secretary of the Aberdeen branch of the National Union of Seamen underlined the distance between them and the inshore men: "They are doing fine on their own. Obviously as seafarers we feel for them and wish them well."[21] Cargill relates a story which suggests that any union would have been hard pressed to obtain a role in the dispute. At a meeting at Peterhead on March 28, a man stood up to declare: "Brothers, you have a great opportunity..." but before he could go on, the cry went up from the hall: "We are'na wanting any Unions here! Pit him oot!" The man was escorted away.[22] Hay himself underlined this basic antipathy, when he argued in 1976 that the only event likely to cause a new blockade would be "if the Government extended the dock labour scheme to the smaller ports", a scheme which was the reason why the inshore men moved from Aberdeen to Peterhead in 1971.[23]

Secondly, the geographical boundary of the dispute was clearly marked. There was no representative of the English inshore industry on the Action Committee, with the exception of North Shields and Grimsby, and no attempt was made to extend the protest any further. By contrast, elaborate procedures were developed to ensure maximum impact for the protest in the ports involved. Tremendous effort went into sorting out lines of communication well in advance of the beginning of the action:

> "Each port had a fleet commander whether it was to be blockaded or not. Ports which were to be blockaded and involved vessels from other ports (for example, Aberdeen was to be blockaded with vessels from Anstruther to Buckie) would have a group controller. It was agreed that the group controller would then be the one to liaise with the AC (Action Committee) in Aberdeen. Local problems would have to be dealt with on the spot, but decisions which were deemed to have over-all significance would be referred back to Aberdeen."[24]

Such clear delimitation of the extent of the action was not visible in France. The 1945 institutions were not able to channel the protest in the way that the Action Committee did. Their interpenetration by the state deprived them of the independence necessary to organise resistance or to provide a focus of loyalty for the fisherman in the local ports. At the same time, their corporatist character made them totally unsuitable for promoting individual rather than collective interests: they echoed the dissatisfactions of the industry as a whole. As a result the CCPM had nothing to do with the organisation of the port protests: in 1980, for example, it only met twice during the dispute, on August 20 and 29. However, at those meetings it presented united industry opposition to both Le Theule's April plan for modernising the deep-sea sector and to his proposals of August 26, directed mainly at the inshore sector. The industry was at one in demanding an increase in the fuel subsidy, drawing attention to the extra money given to the British industry by Walker earlier in the month.[25]

The effects of this lack of delimitation of the protest were twofold. Firstly, the differences between the separate interests inside the industry were concealed in marked contrast to Britain. The most obvious example was the way in which the clash between the crews and owners in the deep-sea industry evolved in 1980. At the beginning of August, it looked like traditional class conflict, with the men locked out by their employers. By the end of the month, the two sides were united around the plea for extra fuel aid, which would enable the owners to avoid a head-on clash with their men. Thus Yves Dhellemes, an important owner in

Concarneau, made his sympathy for the men and its basis very clear:

> "As head of a firm, it is never much fun to have to face up to a strike but today, as owners, we understand the reaction of the men and the decision they have taken to demonstrate in company with their comrades from the ports of the North. For more than two years now the owners have been asking the government for special measures and rates for fuel used by the fishing industry."[26]

Similarly, the strategy of the unions changed. The CGT, in particular, began the month of August by stressing the idea of an alliance between the government and capitalist owners, but it came to see such an appeal as irrelevant in the context of the wider dispute that developed. By late August, increased fuel aid had become "the CGT's sine qua non for the satisfaction of the fishermen's claims."[27] The example provides clear evidence of the way in which the protectionist environment, created by the state, concentrated political demands upon it, rather than encouraging conflict resolution between divergent interests in society.

The second consequence of direct action not being restricted in scope, as it was in Britain, was that the initiative for taking action was at the local level and remained there, even when the dispute assumed a national dimension. The important figure in the 1975 dispute in France was the local leader from Etaples, Bigot, who did not represent a wider constituency, comparable to that of Hay, chairman of the Action Committee, and a man respected throughout the Scottish fishing community. Similarly, in 1980, no major national figures emerged as representatives of the industry in France. In each port, the debate about what to do revolved around the particular interests that were dominant there. This meant that in Boulogne, for example, it was the local CFDT and CGT unions, roughly equal in strength, that were most heavily involved in organising the fishermen, whereas in the inshore ports, such as Port-en-Bessin, individual 'patrons' i.e owners of 'artisanal' boats, channelled the protests. All remained very much isolated from one another and indeed preferred it that way. The idea, put up by the men of Etaples in 1980, that flying pickets move from area to area was rejected as challenging the principle of that autonomy.[28] As a result, it was very difficult to know what was going on anywhere at any one time. And when the fishermen returned to work, they all did so independently of one another, in marked contrast to the British, where central organisation prevailed to the end. Special codewords ('Snowflake' in the Clyde and 'Ben Nevis' in Cromarty) sent the boats home with naval precision on the express order of Willie Hay.

3.2 The limits of protest

The central guidance of representatives of the British industry not only defined the extent of the action but also strongly influenced its character. In the same way, the decentralised nature of the French protests influenced their content as well as their form. In particular, the strength of the representative institutions in Britain enabled the leaders to guide the discontent in a way which proved impossible in France.

It was one of the main decisions taken by the Action Committee at its inception that every effort be made to ensure the minimum level of conflict. It issued clear instructions "confrontation to be avoided at all costs" and "image of the industry must be protected" which were broadly followed by the men in all ports.[29] There was certainly no enthusiasm for pushing illegality too far. Cargill recalls asking why the names of the boats involved in the blockade had not been covered up so as to avoid the possibility of civil action. He received the reply: "oh, that is against the law".[30]

Such reluctance to break the law was more than simply a function of the way the industry was organised. Many fishermen in the North East of Scotland, associated as it is with the religious discipline of the Close Brethren, were very troubled by the prospect of the 1975 protest, while others saw it as a mistaken tactic and cautioned against it when trouble seemed likely to recur in 1980. As Fishing News put it:

> "...the kind of militancy which hits at the public or the country as a whole is hardly the way to win popular support."[31]

However, what happened in 1980-81 itself illustrates the importance of leadership. Atkin, the Chief Executive of the NFFO, set the tone by making it clear that the May 1980 day of protest was not going to be allowed to get out of control:

> "It will be a gentlemanly day of protest. Not illegal but disruptive. We want to show the government the strength of our feelings."[32]

His prediction proved correct but he and his fellow leaders were put under severe pressure to change the tone. There was increasing unrest at boat level, from where it looked as if the industry's representatives were losing touch with their members. A determination developed in the local ports to repeat what had happened in 1975 but against the resistance of the leaders, who were prepared to move towards a national tie-up only "through all the proper machinery of the industry and through legal channels."[33] When there was unofficial action in Grimsby and a sporadic blockade in North East Scotland, they were firmly condemned by the NFFO and SFF respectively. The protesting skippers were very dissatisfied and in Scotland, they passed a vote of no confidence in the SFF leadership and got up a petition to support their position. However, the leaders were not slow to respond: within a week they had won a confidence vote, and the militants' case faded quietly away.[34]

This ability to contain protest was itself not unconnected with the relationship of the leaders with the government. Just as it was important for the government to have some confidence that the leaders were representative of their members and could hold them together, so the leaders needed to feel that the government could itself respond to their promptings. The possibility of militant action was one that the leaders wrote to government about in the early part of 1980 and they pressed the government for help within a particular time limit to avert such action.[35] The government obliged both in March and August 1980, with its packages of aid worth £3m and £14m respectively (cf p 156 above) and was in turn rewarded by a 'responsible' leadership able to limit the protests directed at the state.

What the British case underlines is that the more restrained character of protest in Britain was not the product of some innate quality but rather of the nature of its representative leadership. Without it 1980 might indeed have witnessed something comparable to what occurred in France. Certainly the lack of such leadership was very important in enabling the French protests to develop in the way they did. There was no equivalent mechanism for containing militancy.

In France the CCPM did not limit the character of direct action any more than it defined who should be involved. In its retrospective view on 1980, it took an indulgent attitude towards the blockades of Fos and Antifer, which had aroused such a strong reaction from the government:

> "...if some people in the industry are agreed on the illegality of several of the acts of the fishermen, they are also prepared to acknowledge that those actions were partly justified because of the deep disarray in which French fishing found itself."[36]

Indeed it went further and agreed to pay 100,000FF and 12,000FF to Port-en-Bessin and Sète respectively as a contribution (about 15%) towards the costs that their ships had incurred as a result of their clashes with the navy, thus expressing the collective sympathy of the industry in a similar way to its action in favour of the Breton trawlers the previous year (cf p 231 above).

However, these comments and actions followed the dispute itself. During the protest, the 1945 institutions were a forum for discussion rather than decision-making. They played no role in limiting the protest of the fishermen and there was no effective alternative inside the industry. In the inshore sector, individual ports went their own way and gave up the protest either because they saw no advantage in continuing, as in Brittany, or because they had been so severely treated by the navy, as at Sète and Port-en-Bessin.

In the deep-sea sector, it was the local branches of the unions that were most prominent but their ability to restrict the action was very limited. First of all, they found themselves faced with men who were not willing to accept their recommendations. The deal that was rejected by the men in Boulogne on August 6 was one that both the CFDT and CGT local maritime officers had agreed to with the owners. Secondly, the unions were themselves subject to severe divisions which were reflected at the local level. The general struggle at national level between the reformism of the CFDT and the more resolute approach of the CGT made it increasingly difficult for the local leaders to work together. When a joint agreement was reached in Boulogne with the fish processors to allow them to get fish to their factories, the national headqaurters of the CGT told the local branch to reverse its stance and a total blockade was maintained. In similar vein, the two unions were no longer united when the issue of a return to work was again put to the men on September 5. The local CGT leaders were now obliged to take a maximalist position, maintaining that any settlement had to be based on no reduction in manning levels, while the CFDT unsuccessfully recommended a return to work.[37]

This low level of control by the representative institutions specific to the sector was a peculiarly French problem. Even the TGWU, whose membership was crumbling as the deep-sea industry disappeared, could still claim to speak as the sole representative of the crews of the deep-sea sector and not see its suggestions undermined within a wider political struggle. It worked through its parliamentary links to press its main demand of decasualisation and could hardly be outflanked on the issue. By contrast, the weakness of the French unions in the fishing sector was symptomatic of a wider problem, provoked by the absence of strong representative institutions. Political forces outside the industry were able to manipulate and exacerbate the fishermen's discontents for their own ends, in a way which was distinctly difficult in the British context.

It has already been suggested earlier that the Scottish fishermen were not easily manipulable by the SNP (cf p 203 above). Equally in the present context the Nationalists were not able to influence the direct action of the industry. When a prospective SNP candidate urged the fishermen to consider a blockade of Rotterdam, it was more a question of showing interest in the issue than hoping seriously to bring such a blockade about.[38] Indeed when the blockade had been on, it was the representatives of the industry who determined its limits, with others, such as the SNP, very much on the sidelines.

In France, by contrast, there was much more scope for political parties to seek to mould the protest of the industry. In 1975 the affair was over too quickly for those outside the industry to become involved but in 1980 the situation was very different. The prospect of the Presidential elections of the following Spring gave the opposition parties, in particular, a strong incentive to use the fishermen's case for their own purposes. When the action started to spread in the second week of August, Leroy, a member of the PCF bureau, was able to attack on two fronts. He pointed the finger at the level of the fuel subsidy accorded by the goverment, comparing it with the support just given by the British government to their fishermen. At the same time he could attack the PSF and in particular, Lengagne, the mayor of Boulogne, for their attitude on Europe, arguing that it was the Community which was destroying the French fishing industry.[39] This proved to be the opening shot in a polemic which continued throughout the month. Socialist spokesmen were no less critical of the government but rejected Communist criticism of the party's view on the EEC. As Maire, head of the CFDT, put it:

> "75% of our catch comes from foreign waters. France cannot say 'no' to a European fisheries policy."[40]

This wider political aspect of the conflict served to accentuate the weakness of the representative structure. The political parties had no particular interest in seeking to limit the extent of the protest and were very reluctant to suggest that what was going on was illegal or illegitimate. Even on the Right the prospect of the elections produced anything but unanimity. Those in the centrist UDF saw the villain of the piece as being the Minister, the Gaullist, Le Theule: the deputy for Port-en-Bessin, for example, strongly supported the action.[41] Those in the Gaullist RPR, such as Guermeur, one of the deputies for Finistere, blamed the government in general and Barre, the Prime Minister, in particular, for failing to see that the affair was not a CGT plot to undermine the whole French economy but the natural and spontaneous response of a group in deep economic crisis.[42] Guermeur specifically contrasted what the government had done with the measures taken by the RPR's party leader and presidential candidate, Chirac, when he had been Prime Minister in 1976.

The final comment underlines that there was more than electoral politics involved. What Chirac had done in 1976 and what Le Pensec was to do in 1981 was to underline the character of the fishing industry as a vulnerable producer group that the state was obliged to help in as far as it could. What the Barre government was trying to do in 1980 was to reverse a deeply rooted historical tradition and hence resistance by the fishermen was not seen as subversive of the political order. No newspaper, of whatever political persuasion, was eager to attack the fishermen for committing illegal acts. Le Figaro, for example, a conservative paper, stressed the unpleasant conditions of work and viewed the Boulogne fishermen's case with considerable sympathy.[43]

The contrast with British opinion in 1975 could not be more stark. The Times took a very hard line on the blockade: "The appearance at least must be preserved of not caving in before the latest example of creeping lawlessness in pursuit of sectional grievances."[44] The Guardian, too, was far from happy with what the fishermen were doing. Its Leader commented: "The quickest route to anarchy is to give in to those who break the law."[45] In other words, the lack of the protectionist ethic in Britain provided a much less understanding environment of the reasons for direct action than existed in France. The limits upon such action were therefore very much more pressing in the British than the French environment, limits which the representative leadership in Britain itself acknowledged by their behaviour.

4. Direct action: the state's role

The attitude of the media underlines the point that the industry's stance towards direct action cannot be understood in isolation. The fishermen responded to a wider environment, just as much as they sought to influence that environment's shape. The same point needs to be made in relation to the state, which was not entirely autonomous in the way it reacted to the actions of those in the fishing industry. The way in which in France in 1980, for example, the political parties used the conflict in a pre-electoral situation meant that the Barre government was under a definite pressure: it could not give in without allowing the opposition to make political capital out of what had occurred.[46] However, the government only found itself in this difficulty because of the wider character of relations between the state and the fishing industry. If it had been more sensitive to the traditional pattern of that link, it is highly unlikely that disruption in France in 1980 would have been any greater than it was in Britain, and therefore no opportunity for political manipulation would have occurred. To underline the importance of the state's role in both countries in the context of direct action, the next section will look in turn at the way in which the two states structured their relations with the industry, and the nature of the solution that the different structures produced.

4.1 The structure of relations

After the Left's victory in 1981, Le Pensec, the new Minister for the Sea, attended the CCPM to announce that "the time of scorn and naval battles is past."[47] His attendance was greatly appreciated by a profession which had never previously seen a minister at one of its meetings. This event, and the reaction to it, provides an important element in understanding what it was that made French fishermen feel as desparate as they did in 1975 and 1980.

As we have seen (cf pp 123-9 above), there is no shortage of committees offering the French fishing industry an opportunity to express its views to the government. However, what is also clear is that the main contact is with civil servants from the relevant government departments. The idea of direct links with government ministers was not one that was embodied in the thinking behind the 1945 institutions. In 1975, Cavaillé, the relevant minister, only went back on his initial refusal to meet the protestors when the dispute was assuming national proportions and he felt he had no choice. In 1980 Le Theule proved still more adamant in seeking to avoid involvement in the dispute, arguing that it was one for the "social partners" to resolve themselves.[48] Even towards the end of August, after the Bureau of the CCPM had specifically asked to see him, he remained aloof and insisted on negotiations being conducted through the normal channnels. As a result, when the CCPM met again at the end of the month, its dissatisfaction with the government's proposals could only be voiced directly to Essig, the Director-General of the DGMM: Dubreuil, the Committee's President, was obliged to write to Le Theule to tell him that nobody thought his measures were sufficient "to permit the conclusion of the crisis."[49] As for meeting fishermen from the coast, the Minister left this, too, to his civil servants. When the men from Fécamp, Port-en-Bessin and Cherbourg interrupted their blockade to come to Paris on Saturday, August 23, they, like the CCPM, spoke to Essig and when he failed to satisfy them, they returned to blockade the oil port of Antifer on the Monday.

This refusal to meet the representatives of the industry and the attempt to keep the conflict in separate compartments was not simply a political strategem.[50] There had never been a tradition of direct ministerial involvement in national discussion of fishing policy. The CCPM structure was a classic case of institutionalised consultation as extensive as it was powerless in the context of a major dispute.[51] Moreover, the result of organising consultation in this way was that it was far from clear with whom a minister could have spoken in order to reach some kind of national agreement. The lack of a strong representative tradition in the industry meant that there were very few individuals who could claim to speak for more than a single port. When the fishermen did get to meet Cavaillé in 1975, it was a chaotic affair with as many as 45 local delegates up against the men from the ministry in a battle for increased support. The contrast with Britain is enormous.

One of the important consequences of the move away from port associations to regional or national ones within the British inshore industry was that it created a number of clearly identifiable individuals to whom the government could speak. We have already seen in the previous chapter how these leaders proved critical in Walker's moves to make the consent of the industry a criterion for the settlement of the CFP (p229), and this link was just as important in the more charged atmosphere of February 1975. As soon as the blockades at Immingham and Grimsby had begun, Peart introduced his statement to the Commons with the remark that he had invited representatives of the fishermen to meet him.[52] Even when the response of the industry hardened and the Action Committee was set up, contact was maintained by a compromise which enabled the state to escape the charge of condoning illegality. As Brown, Under-Secretary of State for Scotland, put it, after his discussion with the fishermen on April 2:

> "I took the view that it was inappropriate to meet the Action Committee but I had no objection to the attendance of members of the committee who were on the executive of the Scottish Fishermen's Federation, and two others who were invited as guests of the federation."[53]

More generally, the problem of the blockade was defined in terms of the need for discussion. Brown argued that the very reason that it had occurred was because of "inadequate communication between the Government and the industry."[54] Hence it is not unreasonable to claim that one reason why it did not recur was that ministers were very sensitive to the warnings that they received of growing militancy in their meetings with the representatives of the industry.

By contrast, nothing had really changed in the structure of relations in France between 1975 and 1980. The CCPM was clear that the explosion in the latter year could have been avoided "if the government had been willing to listen more carefully to the many appeals issued by the industry."[55] However, the relationship between the two parties was not organised in such a way as to permit such appeals to have an effect. Instead the government found itself taken by surprise and left with very little room for manoeuvre, a fact which its opponents were not slow to exploit. As we have seen, its initial response was to keep out and hope that the dispute would just 'rot' and collapse, the so-called 'stratégie de pourrissement'. This was, though, a very difficult line to maintain once the dispute became as broad and virulent as it did.

Hence the second reflex which was to take a very firm line in repressing the blockades of Fos and Antifer. Given the special links between the navy and the fishing community since the time of Colbert, these actions were astonishingly maladroit and aroused a correspondingly high level of resentment amongst the whole fishing industry. It is difficult to see what electoral advantage the government could have hoped to gain by such ruthlessness, especially as fuel stocks were not seriously in danger. The inshore boats involved were traditional supporters of the Right, and so could hardly be portrayed as tools of the Leftist unions.[56] More to the point, the strategy revealed the way in which other options could be seen not to exist once the French state was faced with centrifugal pressures strong enough to undermine the generally high level of order within this particular sector of society.

A third strategy was developed by this government to cope with the situation. It consisted in offering piecemeal concessions, such as those announced by Le Theule on August 26. Whereas these suggestions might have been productive at an earlier stage, by the end of August opinion in the industry had hardened considerably. It did not perceive the Minister's proposals as concessions, even though they were certainly more significant in financial terms than anything suggested by the British government in 1975 in its successful attempt to defuse the blockade. Such was the cost of the more brittle, less flexible structure of relations that existed in France.

4.2 The nature of the solution

As the last paragraph suggests, the structure of relations and the solution it produces cannot be kept rigorously apart. Indeed it is arguable that in the British context, the structure was itself one of the solutions. When Hay called off the blockade, he gave two reasons: firstly, 75% of the demands had been met; secondly, there was a genuine belief that the Government meant business.[57] The very vague content of the statement by Brown, the minister who negotiated the ending of the blockade, made the first of these look decidedly optimistic. As Fishing News put it, it was "Back to Sea on a Promise."[58] However, this very fact underlined the importance of the second reason given. The Scottish inshore industry, in particular, felt that it was now being considered more seriously and that the UK fishing policy would no longer be geared predominantly to the demands of the BTF and STF.

Subsequently, the possibility of direct action was something that Walker, in particular, guarded against with great political skill. As soon as he announced the aid package of August 1980, the threat of unofficial action by the Scots petered out. Similarly, when the boats throughout Britain tied up in February 1981, he responded quickly by saying that he was bringing forward a review of the subsidy scheme, a move which the representatives of the industry could point to as indicating the success of their pressure tactics and which the government could present as successfully bringing the action of the fishermen to an end. At the same time, the particular complaint of the fishermen against the large imports coming into the country was one that Walker dealt with in a way designed to defuse the situation rather than provide a specific solution. Firstly, he put the burden of argument on the industry, by saying that he would do something, if it could prove that 'dumping' was to blame for the low earnings of the fleet. Secondly, he set up a committee of enquiry with six representatives of the industry under the chairmanship of the Fisheries Secretary at MAFF, Mason.[59] The report itself emerged six months later when the situation was much less tense and concluded that the basic problem was the high value of sterling rather than unfair practices by exporters to Britain.[60] Indeed as the year progressed, controls on imports were quietly relaxed as their adverse effect on processors became clear. In this way politically adept solutions were devised for potentially explosive situations.

The way in which similar situations in France were treated was quite different. Rather than seek compromise, the French state was torn between surrender and technocratic repression. The first option was taken in 1975. Chirac indicated to his Minister that he should concede to the fishermen's demands and the question at issue was simply the extent of the concessions that would be made to calm the fishermen down. Thus on top of the 8MF agreed to on the morning of February 20, the weary government negotiators agreed to a further 3MF in the afternoon.[61] It was not a question of putting off a response to a later date until a committee had reported or the shape of an aid package had been agreed. Nor was this the question in 1980 when the second option proved more acceptable to the government. Everything that was done confirmed the idea of the technocratic character of the French state, where technical competence and a sense of public service amongst administrators is combined with a woeful inability to understand and respond to the desires and expectations of the objects of government, the 'administrés'.[62]

At the level of individuals, the nature of the state response was epitomised by the role played by Essig, the Director-General of the DGMM. He had had a brilliant career with the Regional Planning Agency (DATAR) and been drafted into the Ministry of Transport in September 1978. He was not a man with a feel for the difficulties of the fishermen, however well he could analyse them, and his association with the events of August 1980 led to his being moved on at the end of the year. Indeed one of the moves of the new Socialist government to correct this bias was to appoint as Director of the Fisheries Directorate, Proust, a man with long experience as a maritime administrator on the ground, and very much respected by the fishing fraternity.[63]

However, it was not simply a question of personalities. It was also a matter of the way in which arguments were openly used to try to suppress the validity of the fishermen's case. In no sense did Le Theule see himself as their champion against other interests, rather he called on facts and figures as if he were some kind of neutral arbiter. Thus in the second week of August, he sought to sway opinion against the idea that the financial situation in the industry was as bad as was claimed by quoting figures for the average earnings of a trawler skipper, a chief engineer and a member of the crew. He suggested, for example, that a crewman could expect to earn 117,000FF per year, a total which provoked a reply from the owners that in reality, the figure was unlikely to exceed 70,000FF and that in any case, the intervention only served to exacerbate the situation.[64]

A second example of a technician's attack on the fishermen was the use of the Community 'card' as an argument against according a higher fuel subsidy. The EEC Commission sent a letter to France on August 12 threatening to take her to the ECJ, the Luxembourg court, for unfair competition in view of the fuel subsidy she was already giving. This was to provide ideal ammunition for both Le Theule and Barre, who noted that French fishermen were in any case paying less for fuel than almost all other European fishermen. However, to argue that EEC law left the government with no choice would be quite unrealistic. As Eisenhammer puts it:

> "These official fuel prices mask the fact that most countries practice hidden subsidies of some sort. The surreptitious manner in which the French government gave massive aid to the Saint-Nazaire shipyards, in contradiction of Brussels' regulations, and the fact that the newly-elected Mauroy government in 1981 covertly doubled the fuel subsidy, shows that where there is a will there is a way."[65]

In a similar way, the solutions that were proposed as well as those that were excluded reflected technical rather than political competence. At a Paris meeting to discuss the case of the Boulogne deep-sea crews on September 2, a compromise was put forward by the administration involving the rotation of fishermen rather than any redundancy. Those staying on shore would receive a reduced wage supported by the National Employment Fund. However, as was remarked at the time, there was no attempt to 'sell' this proposal. Ministers failed to utter a word of sympathy or understanding to make the ideas palatable after such a prolonged and bitter dispute; it was a case of 'too little, too late'.[66]

The contrasting British and French states' responses to direct action were not only dictated by considerations of how best to deal with an interest within society. There was also the issue of the broader way in which the problems of that industry should be conceived. As we have already seen in Chapter Five (pp154-63) the philosophies of economic intervention that applied in the two countries were very different. In particular, there was a contrast between a French protectionist tradition which dictated a thorough attempt to shape the fishing industry, and a British liberal tradition which accepted a need for intervention without depriving the industry of the main role in determining its future. This difference was important in the context of the tension generated by direct action in that it established different expectations of what the state should do.

In the French case, the unwillingness of the Barre government to contemplate an increased fuel subsidy suggested that it was considering imposing an important change on the shape of the fishing industry. Barre argued that to concede on the fuel issue would not only open the way to similar demands from other interests, such as taxis, but also enable fishermen to avoid the 'real costs' of energy, which he was eager all should face.[67] The position was not solely determined by the blockade: later in the year, Le Theule's successor, Hoeffel, remained opposed to fuel aid, despite his generally more cooperative stance.[68]

Such a liberal leaning opened the government to the charge that it no longer wanted to protect the industry as all governments had in the past. In the documents of the 8th Plan, it was pointed out that without the 'political will' there could be

8,000 direct and 40-50,000 indirect job losses in the industry by 1985. So successful was the CGT in arguing that this would happen because the government lacked the necessary will that by the end of the month, foreign commentators of liberal economic views had picked up the argument and accepted it.[69] Hence it was not just a question of an authoritarian style on the part of the government but also the fact that its ideas for solving the crisis ran contrary to a traditional pattern of economic support. In this way, the political and economic strands of the French state tradition can be seen to merge in an explanation of the response of French governments to direct action.

The British example underlines a similar combination of the different strands of tradition. Just as politically the stress was on concession and compromise when direct action was taken or threatened, so economically there was a reliance on short-term *ad hoc* arrangements, designed to appease but reflecting a marked reluctance to run counter to a liberal interpretation of intervention. For this reason, British governments could use similar arguments to the French about the knock-on effects of a fuel subsidy without exacerbating discontent. In the same way, they could avoid the pressure for import controls, which had a clear effect in France. Walker's response in 1981 to the pressure for import controls in appointing a committee was not just a clever way of sidestepping the problem but also reflected a long-standing economic tradition, which the threat of direct action by fishermen could not hope to undermine.

5. Conclusion

The present chapter has come to two main conclusions. First, it has confirmed our initial expectations that the severity of the changes in the issue area would induce a tendency towards direct action in both states but that the tendency would be more marked in France given its particular traditions. In so doing, it has underlined distinctive features of the character of the industry and the state in the two countries. The French industry has been seen to conform to a decentralised pattern of interest group behaviour with leadership coming from behind,[70] while in Britain representation was concentrated in the hands of powerful individuals, able to influence significant sections of the industry. The contrast was one which the two states had helped to bring about but which also influenced their behaviour in that British governments found themselves with a reliable interlocutor in the search for compromise, while French governments were driven to choose between repression and surrender in controlling centrifugal pressures.

Secondly, the chapter has underlined the link between this perspective and the previous ones. It has stressed the way in which the same divisions within the British industry which were so important in the mediatory perspective in preventing a united view from emerging were critical in limiting the extent of direct action, given the strong leadership of each section of the industry. By contrast the lesser degree of open articulation of differences in the French mediatory context served both to harden opposition to the state and to limit the possibilities of conflict resolution when tension increased. Similarly, the state's response to direct action has been seen to be linked to the interventionist perspective. The British liberal economic tradition rendered the defusing of conflict more straightforward than it was in France where protectionism generated more specific expectations as to the way in which the grievances of the fishing industry should be resolved.

8 The self-help perspective

1. Introduction

The fourth and last of the chapters organised around a perspective will consider the efforts of the fishing industries of Britain and France to manage their own economic environment. It adopts therefore a 'bottom-up' view of the link between state and society which complements the 'top-down' approach adopted in Chapter Five. As with the previous three perspectives, this self-help perspective is linked to two claims as to how we might expect the relationship between the fishing industry and government to develop (cf p 31 above). The first of these is that "the industry would respond to change by intensifying its efforts to help itself". We will find that this did occur but that the industry's efforts cannot be considered in isolation. They were heavily influenced by the attitude of the state towards them. Hence we will also find confirmation for our second claim that "the difference between the character of French and British state intervention in the economy would make such self-help efforts much less individualistic in France than they were in Britain".

The chapter is organised into three parts. The first will suggest that the fishing industry has always been marked by a strong clash between the individual and collective interests of catchers. Before the 1970s neither country had much success in reconciling this clash of interests in the inshore sector but in both countries, the deep-sea companies were able to develop structures of joint self-management. In France these structures were heavily influenced by the state, in Britain they were the product of mutual agreement amongst the companies. The second section will argue that these contrasting views of self-help were reflected in the development of EEC producer organisations. In France there was a shared determination on the part of state and industry alike to use them in order to develop a common producer interest. In Britain the EEC structure was not used in the same way but generated a fragmented pattern of activity within the industry, subject to no significant coordination by the state. The third section will claim that the growth of producer power in Britain was blocked by the existence of a broader set of trading and processing interests, closely entwined with those of the catchers. No British government was prepared to disentangle these various interests or to accord clear priority to the interests of fishermen as national producers. In France, by contrast, the interlocking of interests was less marked and there was strong pressure to protect the national producer interest. The result was that self-help efforts in Britain lacked the collective shape they had in France.

2. Individual and collective interests

The fierce individualism of those involved in fishing is legendary. As one writer put it: "every fisherman is a poacher and a smuggler at heart."[1] The result is that there has been tremendous competition between fishermen with each man determined to earn more than his neighbour on his return to port. In the British deep-sea industry, this competition was institutionalised and legitimated by the annual award of what was known as the 'Silver Cod' trophy to the trawler captain who made the largest catch in a year. The winner could expect quite a contest from his fellow captains, even those within the same firm, the following year, and the tactics used were anything but gentlemanly. A favourite trick was the placing of a sack over the cod end, i.e the rear of the net so that fish could not escape, whatever the net's mesh size, and the vessel's catch could be increased accordingly. More dangerously, a vessel might give the impression of lowering its nets over ground known to be strewn with 'fasteners' or obstacles in the hope that a competing captain would fish in that area and damage or lose his trawling equipment.[2]

Within the inshore industry, the attraction of making a bigger catch than one's competitor has encouraged fishermen to risk contravening legal rules. Despite the widespread belief in Britain that French ships were landing herring illegally in France, as illustrated by the March 1980 TV documentary (cf p 93 above), it was not beyond Grimsby inshore captains to sail to Denmark to land herring as a way of escaping British controls and making money at the same time.[3] It was a classic example of the 'free-rider' problem which the individualism of the industry exacerbates. Why should fishermen risk their economic survival if they cannot be sure that others will adhere to the rules?

Such competition has created an important barrier between the industry and the state. Fishermen have been reluctant to reveal too much about practices that can bring them economic advantage. Thus the state structure is well aware both of the unsatisfactory nature of much of the information it receives as well as its dependence upon the industry for that information. As the French Fisheries Directorate put it in 1976:

> "Unfortunately, very many captains and owners have still not understood that the future of their industry is linked to the quality of the information that they ought to supply and there is reason to fear that it will not be possible to defend their rights effectively in the years to come."[4]

Similar kinds of worry emerged during the Trade and Industry Sub-Committee's discussion of fish dumping. The Director of Research at MAFF made it clear that the government is obliged to accept what the industry tells it and can only hope to obtain very rough estimates made:

> "either by interrogation of skippers or mates on landing or, in some cases, observers have been put aboard the vessels, but the number of observers one can put on vessels is clearly few."[5]

However the same secretiveness which limits the information given to governments has also limited the development of the collective rather than individual economic interests of fishermen.

Fishermen are anything but eager to share with other fishermen the information which helps to determine their individual wealth. For example, in order to maintain their own secret knowledge of where and when to make the best catches, they will frequently switch their radios off so as to avoid giving away their position away and they will never wish to indicate how much they have caught in any particular place.[6] Yet the common organisation of the catches made by fishermen is one way by which they can avoid gluts in the market, and the consequent fall in price which affects all of them adversely and can only benefit retailers and consumers. Agreement on collective action could make all catchers better off. But the strength of individualist ideology hindered such an agreement and adequate structures of self-management. The nature of the structures that did come into being will be considered in turn for the inshore and deep-sea industries.

2.1 Inshore industry

Before the 1970s, the cooperative was the basic mechanism for self-management that existed in the inshore sector. The cooperative offered a forum for fishermen to take economic decisions, while their local association (in Britain) or local committee (in France) provided a mechanism for representing their views to the authorities. Through the cooperative, the inshore industry could hope to act directly to improve its economic position.

In both countries, the idea of encouraging the industry to come together and organise itself in this way goes back to the early years of this century.[7] In Britain, as we have seen (p130 above), the government established the FOS back in 1914 'to foster the propagation of cooperative principles' within England and Wales (in Scotland, the cooperative movement only began between 1940 and 1945). An annual Treasury grant (amounting in 1972, for example, to £14,000) was given to the FOS to assist it in its task and to supplement the contribution that it received from its member cooperatives. In France, too, the state took a lead, though establishing a much more direct supervisory role. A law of 1901 gave maritime administrators the right to attend and exercise a veto at meetings of the cooperatives, and in 1913, a common bank for the inshore industry, the Crédit Maritime Mutuel, was established under state decree. This bank was given the job of receiving the membership claims of cooperatives, agreeing to their statutes and guaranteeing them credit.

Since that time the system has spread to include ten regional banks in mainland France and three in the DOM-TOM. These bodies along with the cooperatives have come together to form the Confédération des Organismes de Crédit Maritime Mutuel (COCMM), which gives a national sounding board to the inshore cooperative movement. In Britain, by contrast, the FOS remained very much a regional phenomenon, concentrated in the South West of England, and undermined as the spokesman of inshore men by the growth of the NFFO in most of the rest of England and Wales.

Despite these long-standing arrangements and despite their diverse forms, neither country was notably successful in generating a firm cooperative structure. In France, there was an increased interest in cooperatives in the 1960s but by 1970, 50 of the 69 coops were still restricted to dealing with the common purchase of supplies, such as rope, fuel and nets, and only ten were concerned with the marketing of fish once caught. By contrast, in Britain, including Scotland, only 25 of a total of 125 in the middle of the 1970s were supply cooperatives and only 54 were marketing cooperatives, the remainder being representative associations attached to the FOS as non-trading societies. However, the fact that members were not obliged to deliver their entire catch to the cooperative for sale and that membership, particularly in Scotland, was not high meant that the marketing cooperatives were not very successful. No more than half of the inshore production in England and Wales and as little as 10% of Scottish landings were marketed by cooperatives in the early 1970s.

In both countries there was a constant problem of obtaining the right kind of people to run the cooperative. For long periods of time the fishermen themselves would be at sea and their level of involvement would extend little beyond a small financial contribution to its expenses. The management of the cooperative would necessarily be devolved to someone who did not go to sea, and the financial rewards offered to that person could not be high. The typical kind of situation that arose as a result was described by the Chairman of the Hastings' Fisherman's Association in his evidence to the Trade and Industry Sub-Committee:

> "Up until about 12 years ago we used to have a fishermen's cooperative. There was not enough money going into the cooperative society to get anybody to run it full-time and we had a retired fisherman running it. We could not afford to pay him a lot of money and when he died we had to pack up the cooperative."[8]

There were some exceptions to this general rule in both countries, where local circumstances proved to be particularly favourable. As early as 1945, at Eyemouth in Scotland, for example, the Presbyterian minister played an important role in establishing a coop which by 1972 had 400 members and an annual turnover of over £1 million. Moreover, it took on a remarkably wide range of tasks, going well beyond the provision of supplies and basic marketing. It established deep-freeze and storage facilities; it worked out independent withdrawal price systems; and it organised market intelligence on the level of prices in other ports. Similarly, at Etaples, near Boulogne, the cohesion of the fishermen which was to drive them to Paris to protest in 1975 (cf p239 above) enabled them to form a cooperative in the mid-60s, which came to work for virtually all the 'artisans' in the area. Like the Eyemouth coop, it set up a freezing capacity so as to cope with gluts on the market and could exercise sufficient authority over its members to organise limits on the levels of catch.

However, the environment of the cooperatives in both countries has been transformed by the EEC structure of producer organisations, agreed to in 1970. This structure offered an opportunity for the inshore industry to manage its economic welfare in a much more direct way than it had done in the past. The result was that some cooperatives like Etaples, became POs, aiming to bolster first hand sales of fish, while others continued to exist as separate entities but in a new and close relationship with the POs in the search for outlets for products. The character of the balance between individual and collective interests was radically altered in the process.

2.2 Deep-sea industry

The difficulty of managing conflicts between individual catchers proved less severe in the deep-sea industry than in the inshore industry because of the existence of shore-based companies. Whatever the conflicts between captains on the fishing grounds, the companies in both countries came to recognise, as early as the 1960s, that it was in their collective interest to control the competition between themselves. Above all, they recognised the importance of the need to organise the link between the catching and the selling of fish. They devised mechanisms to avoid one of the constant difficulties of the inshore sector, namely that vessels tend to leave port together and to return together, thus provoking gluts on the market and a general lowering of prices. Not only would owners phase the return of their boats so that they did not land their catch together, but they also supplied information in advance to the merchants and processors as to what fish and in what quantities a ship would be returning with.

However, within this general framework of increased cooperation, the two countries differed markedly in the way in which the pursuit of collective interests was organised. The French industry invited the state to establish the terms of cooperation and accepted a degree of direction determined by public law. In Britain the arrangements were the product of informal agreement amongst the companies and imposed a much more limited set of obligations on those involved. It was a classic example of the contrast between interest accommodation in a state-led society and a society-led state.

As we have seen in Chapter Four, the French industry had been involved since 1945 in a degree of collective economic management, through the Interprofessional Committees. However, in the early 1960s, the feeling grew that these committees were not enough and so the industry pressed the administration to create new, more wide-ranging institutions. Thus from 1965, the structure of the 1945 Decree was supplemented by three Regional Funds for the Organisation of the Market (FROMs), one for the North region, a second for Brittany and a third for the South West. Each was run by a governing body with representatives from the industry and the state.

All three FROMs were given significant powers to intervene in the market and to implement a system of legally enforceable withdrawal prices. If fish could not be sold above the withdrawal price the company concerned was obliged to place its fish on a 'second market', where it would be bought for conversion into fishmeal. It would then receive from the FROM the difference between the price obtained on the 'second market' and the guaranteed withdrawal price. To ensure this level of protection, all companies were obliged to belong to the FROM in their area. They also had to accept instructions from the FROM such as changing a ship's course, if the market it was coming to was saturated. To ensure adequate financing, alongside the compulsory contributions of the industry, the state provided substantial sums: in the case of FROM Nord, for example, 42% of the budget in 1965, 25% in 1966 and 50% in 1967.[9] in this way, the FROM were able not only to pay their members, if prices fell, but also to offer special premiums to processors if they were prepared to accept particular quantities of fish in advance.

The British arrangements were very different. First of all, there was no question of the state being involved institutionally or financially: consultation with the WFA was the limit of involvement by those outside the industry. Secondly, the system that was established, known as the Distant Water Vessels Development Scheme, set minimum prices below which the members of the scheme would not sell their product: it did not establish a system of price guarantees should prices fall below the minimum levels. What was designed to keep prices up (and succeeded in doing so between 1966 and 1973) was the mutual agreement of the owners not to undercut their competitors rather than the availability of finance for sales under the minimum prices set by the owners. Finally, the fact that the arrangement was not embodied in law meant that it could be attacked in the courts for distorting the market. The BTF was obliged to defend the scheme before the Restrictive Practices Court in 1966 and was successful. But in Scotland, the equivalent Court issued an order in 1970 preventing the STF from operating any scheme designed to limit the mode of operation of a trawler firm.[10] Only in 1975 was this order rescinded as a result of the application of EEC legislation on producer organisations.[11]

This last comment underlines how important the development of EEC policy was for the industry. The PO system opened up the opportunity for part of the financial burden of withdrawal and guarantee prices to be undertaken by the Community. However, the catching sector was not unanimous in perceiving the usefulness of such support for their efforts at self-management. The FROMs in France moved rapidly to obtain status as POs, but in Britain the process was much slower. The first operation to withdraw fish from the British market did not take place until the middle of 1975.[12] The reasons for this need to be sought in the different ways in which the two industries responded to the possibility of EEC support for their activity.

3. EEC producer organisations

In Chapter Three (pp 96-114) we saw that Britain and France had consistently different attitudes towards the CFP. France's major concern was to ensure maximum protection for her fishermen whereas Britain wanted as large an area of exclusive access and as substantial a level of quotas as possible. Hence the creation of the system of producer organisations in the fisheries sector did not have the same meaning for the two states. Though Britain was just as much bound by this part of the 'acquis communautaire' as any other, for her and her fishing industry it was a far less significant aspect than it was for France and her industry. For the French the existence of the PO framework was a critical element in a policy they wished to use to protect the interests of their fishermen. For the British the need for such protection was not perceived in the same light, and POs received a correspondingly smaller level of attention.

3.1 Historical development

In the early 1970s most British concern was expressed about the idea of fishing 'up to the beaches', as the 'equal access' provision suggested would occur after the ending of the derogations agreed under the Treaty of Accession. In as far as there was discussion about the CFP more generally, the marketing arrangements were not commented upon very favourably. The idea of a unified pricing structure was seen as very inflexible and likely to lead to major distortion in the market. The fact that it might assist fishermen as a group was not necessarily perceived as an advantage. Laing of the BTF noted a possible parallel with the difficulties of the Common Agricultural Policy (CAP):

> "...the policy is explicitly directed towards the protection if not the promotion of the interests of producers rather than consumers. If EEC agricultural policy provides any precedents, then the extent to which or at any rate the pace at which rationalisation of the fishing industry will proceed must be a matter of doubt."[13]

In other words, there is little evidence that the industry itself in Britain was particularly interested in having to deal with a further layer of bureaucracy, at a time (before 1975) when it was doing very well without such interference in its affairs.

In France, by contrast, there was every interest in the potential of producer organisations. As already indicated (p 101 above), French negotiators were concerned to obtain as much protection as possible in the face of the impending dismantling of all barriers to trade in fish products between Community members in 1971. They were determined to ensure 'an equal chance' for French fishermen in the face of the freeing of the movement of the factors of production. The fact that consumers had traditionally been obliged to pay higher prices than in other countries was seen as a necessary effect of guaranteeing producer income.[14]

Ironically, the French industry was anything but happy with the results of the negotiations. As the CCPM commented:

> "The Community idea of producer organisations is, in our eyes, unrealistic, because these organisations are neither obligatory nor interprofessional."[15]

In other words, the industry felt that the PO system could only work if all fishermen in a particular area were obliged to belong, allowing no opportunity for individuals to sell outside the official network at a lower price. Although French pressure in the EEC resulted in three years being fixed as a minimum time for participation in a PO, it proved impossible to get all EEC members to agree to the idea of compulsory membership.[16]

Despite continuing complaints it was recognised in France that the POs could play a very major role in helping the inshore sector, in particular, to overcome the inadequacies of its old cooperative structures and that they could come to have an increasing role in controlling the market. Such potential was perceived much more slowly in Britain. However, in both countries it was the events of 1975 that gave further impetus to the debate about POs.

1975 was, as we have seen, a very difficult year for the fishing industry in Britain and France. It prompted higher government intervention, it increased the volume of complaint from fishermen and it produced unprecedented levels of direct action. What it also did was to encourage a tendency for the industry to seek to help and organise itself. In France, in particular, there was an increased awareness of the possibilities of using the PO structure to overcome internal infighting. On the Mediterranean coast alone, seven new POs were set up in the first half of the year, when there had been none before.[17] In April, 18 POs, inshore and deep-sea, from throughout France decided to combine to form a National Association of Producer Organisations (ANOP). At their inaugural meeting, they were given a symbolic speech of support by the infamous head of the vegetable cooperative of St-Pol-de-Leon, Gourvennec, whose members had used their combined wealth to set up Brittany Ferries.[18]

In Britain, too, it was recognised that POs offered a new way forward. In July, the SFO, the largest PO in Britain, was able to fix herring and white fish prices. This was possible because the legal ban on the setting of minimum prices in Scotland had been lifted and its members could now realistically start to encourage other fishermen to join. However, it was by no means easy to get fishermen to conceive of their interests in collective rather than individual terms. Membership remained low and awareness of the separate interests of the different sections of the industry meant that it proved very difficult to create an equivalent of ANOP. In February 1976, an association of inshore POs was set up. But it excluded the major Scottish inshore interest, the SFO, and consequently foundered to be replaced later in the year by the United Kingdom Association of Fish Producer Organisations (UKAFPO).[19] As we shall see, it too had a difficult career.

Despite the difficulties in Britain, the years that followed produced an important growth in the range and activity of POs in both countries. By 1982, the coverage of the coastlines was almost complete: there were 14 in Britain and 21 in France (including one for the tuna fleet based at Dakar in Senegal). Moreover, they came to be involved with each other in a wider arena. In November 1976, at the instigation of ANOP, POs from all over the EEC met for the first time in Brussels. In June 1980 a European Association of Producer Organisations (EAPO) was established to act as a lobby for improved price levels for fish products and a tilting of Community policy in favour of organised producers.[20] Thus the POs were becoming more than economic agents. They were moving into the political arena of obtaining resources, the task traditionally undertaken by the representative associations, such as the SFF and the NFFO in Britain, the CCPM structure in France and 'Europêche' at the European level. In this way, the difference between self-help and the more traditional notions of an interest influencing government became blurred.

However, the blurring of the distinction does not alter the fact that the producer organisations retained a distinctive role. They were concerned to act directly on the market in order to assure a 'fair income to producers' and were not simply interested in persuading government to improve that income. At the same time, they were not just agents of the state or the Community but retained an important degree of autonomy. They could only be established 'on the producers' own initiative' and they were free to decide whether to apply EEC prices or autonomous prices, though forfeiting Community support if they chose the latter. This did not mean, however, that they operated in similar ways. The contrast between their role in Britain and France after 1975 was very marked and illustrates the importance of the state's role in channelling the self-help efforts of the fishing industry.

3.2 The role of the state

What happened in France to the PO structure cannot be understood without recognising the role of FIOM. In Chapter Five, the importance of this body in encouraging experimental voyages was stressed (p177-89 above). However, it also became the effective patron of the French POs and their link with the administration. By 1982 over half of its budget (63MF) was being spent to help the POs in their efforts to intervene on the market and to complement the funds coming from the Community, while the POs themselves had come to turn instinctively to FIOM as the guardian of their interests.[21]

The close relationship betwen FIOM and the POs was particularly clear in the operation of a system of national withdrawal prices. As the Community-guaranteed withdrawal prices only applied to 13 species and France markets over 40, there was a clear difficulty over how to deal with non-Community species. The system that evolved from 1976 was for ANOP to agree to, and present, to FIOM prices for these non-Community fish. Once agreement was reached on these prices, they were presented to the merchants to allow for further bargaining. From then on, FIOM would guarantee financial support, if prices dropped below the agreed withdrawal price level. The mechanism of guarantees was based on a close tie with all the POs. At the beginning of each year, FIOM would establish a total value of the 'drawing rights' or 'credits for market support' to which each PO was entitled. During the year, the PO would itself pay for withdrawals and receive 45% to 60% of the money back from its FIOM allowance. If it then reached the limit of its credit, the Fund's administrators could acknowledge a particular difficulty and grant more money or tell the PO to find the money out of its own resources.

In 1980 this basic support mechanism was supplemented by the creation of a system of 'target prices', intended to compensate for loss of earnings. POs were allowed to choose 10 species and for each of the species the average price that obtained in the previous year was calculated. This figure was then increased by 10% to produce an 'target price' for the current year. Where the actual price received by the fishermen was lower than the 'target' one, FIOM made a contribution of 50% of the difference to cushion the blow of decreased income, with the remainder coming from the PO's own funds. The scheme proved popular with the POs but was also very expensive. As a result, the number of species that each PO could choose was reducd to three in 1981 and then zero in 1982. However, the very existence of the scheme underlines the closeness of the link between FIOM and the POs. British POs looked at the arrangements with an envy as large as their chances of obtaining a similar scheme were small.[22]

In addition to its role in supporting prices, FIOM also made tentative steps towards obtaining a larger level of control over the market beyond the first-hand sale. It encouraged supply contracts between producers and processors, offering the latter a small financial incentive of 20 centimes per kilo for such deals. It was anything but easy to get such arrangements, with producers inevitably seeking a high price and processors a low price, neither eager to settle for a fixed price. However, what was clear was that POs looked to FIOM to help them in this domain: some even went so far as to want it to be obliged to take all surpluses and to get rid of them as best as it could, so that the fishermen would cease to have to worry about what happened after first sale.[23]

In other words, in the French context, the self-help reflex was strongly conditioned by the knowledge that the state was supporting the producers' initiatives and offering important financial backing. This was not the case in Britain, where there was no equivalent to FIOM and the only special assistance for POs came in the £2m allocated to them in March 1980. None of the subsequent aid packages contained money specifically allocated to the POs. In part, this can be understood in terms of the different British market structure: the main British catch was covered by the Community withdrawal system, unlike the French. Equally, the government was reluctant to contravene EEC rules concerning the level of support to be given to POs. More important, however, was the British aversion to interference in the economic arena and its greater enthusiasm for competition within the market.

It has already been indicated that the French fishing industry was disappointed that the Community arrangements did not conform to their own consultative structure and make membership of POs compulsory. Hence there was constant pressure throughout the 1970s for the extension of discipline to non-members to prevent fishermen picking and choosing whether or not to sell through POs. The 1983 EEC deal did make it possible for members and non-members of a PO to be differentiated, the former receiving up to 100% of the withdrawal price, the lattter restricted to 60%, but there were major British doubts about any further discipline. Lord Mansfield, the Scottish Fisheries Minister, had stated clearly the government's worries about how much compulsion should be exercised, referring to the:

> "fundamental question of whether it would be right to seek to impose on those fishermen who have chosen, for whatever reason, not to join a producer organisation an obligation to conduct their affairs in certain, as yet unspecified respects, as if they were members of a PO."[24]

In such 'voluntarist' circumstances, it was hardly surprising that British POs were not able to develop the same kind of relationship with the state that their French colleagues could.

However, it was not simply a question of an unwillingness to intervene in private choices. There was also a strong feeling - particularly strong under a Conservative government - that the acknowledged problem of an uncertain market demanded a higher level of entrepreneurial skills, not the development of producer power under state supervision. Walker, the MAFF minister, commissioned a marketing report which appeared in July 1981 and called on fishermen to stop being 'quarelling gamblers' and to cooperate.[25] However, the authors of the report did not see POs as the way to further cooperation. Rather the stress was on the need for greater fish promotion through the selling of new species and the use of catchy new brand names - MacFish was suggested! The report provoked a considerable amount of comment and debate, ending in a major conference in November, but all this activity served to underline a general belief that market conditions could be improved by simply pointing to the benefits of cooperation. Why this should be seen to be enough to bring about cooperation, when any individual could hope to benefit from it without making a contribution remained obscure.

The general tendency of government policy was confirmed by the nature of the SFIA, which replaced the WFA and HIB in October 1981. Far from acting as the financial sponsor of the POs in Britain, the new Authority devoted its first efforts to encouraging the purchase of fish. Its method was a large scale marketing campaign transported round Britain by train. Moreover, its own structure made it difficult to envisage it ever becoming more directly involved in the allocation of resources to the industry. The presence of eight industry representatives on the Board of twelve brought the Authority closer to fishermen than its predecessors but it also made it more difficult for it to take decisions benefitting one group more than another.[26] In other words, the notion of a collective catcher interest remained much less developed in Britain than it was in France.

3.3 The societal response

The behaviour of the producer organisations themselves was a reflection of the state's attitude towards them. In the French context, they flourished; in the British context, they faced a constant up-hill battle. There was an important psychological element in that in France the POs felt strongly that they were operating within an officially-approved environment. As early as 1976 (cf p 143) the French state gave exclusive recognition to ANOP, while five years later there was still a belief that the British had failed to match the unity of the French. As the head of the Bridlington PO put it:

> "We desparately need to breathe some life into the UK Association of Fish Producer Organisations so that it can become a vital forum for mutual cooperation and benefit."[27]

Without the kind of supervision offered by FIOM it was hard to see how the environment for such unity could come about: the divergent interests within the industry were too great.

The result of this difference in environment was that the French POs found it easier to act both in concert and alone.

Already in 1976 an agreement was reached in ANOP that the PO to which a boat belonged would be responsible for the payment of guaranteed prices and the landing taxes of a ship, even if the ship put its catch ashore at a port other than its home port.[28] In Britain, by contrast, UKAFPO was never able satisfactorily to cope with the conflicts that arose between the 'nomad' ships of Scotland and Humberside, who sometimes fished for mackerel and the small-scale indigenous fishermen of the South West coast. The difficulty may have been greater in the British context but the institutions were certainly not designed for the resolution of conflict.

The belief that the French state considered membership of POs important - government aids to investment, for example, were less favourable for non-members - increased members' willingness to hand over responsibilities to their individual POs. This was reflected in the fact that French POs were more easily able to levy higher contributions than their British counterparts. For example, members of the SFO paid half a per cent of the gross value of their catch to PO funds, whereas members of FROM Nord were estimated to pay almost 2.2% of their gross earnings. Though the latter figure was particularly high, the average in France was reckoned to be around 2% of the value of landings.[29]

This does not mean that British POs were necessarily poor. We have already seen that the SFO had an income in 1979 of £300,000 (p 140 above) but even earnings at this level were not sufficient to cope with all the problems that were faced by the industry during the period of this study. When, for example, prices fell at the auctions, many British POs felt that the Official Withdrawal Prices (OWP) of the Community were inadequate to guarantee income levels. So they decided to forego EEC support and to apply Autonomous Withdrawal Prices (AWP). However, this proved very expensive indeed, as the SFO discovered. In 1979 it paid out such large sums of money (£325,000) that it was obliged to stop to avoid going bankrupt.[30] Throughout Britain the pattern was similar with different POs applying different AWPs. What was lacking was any degree of coordination like that which which existed in France. There, even the single PO which did not belong to ANOP - that of Etaples - applied the national withdrawal prices agreed within the FIOM framework.

It is true that the British market is not directly comparable with the French: the former is dominated by large quantities of a smaller number of species. However, there was also a difference of attitude. The French POs were not content simply to apply the withdrawal system, whether official or autonomous, because they recognised that they too would go bankrupt if they did. Within the inshore sector, in particular, an important link was established between the POs - forbidden under EEC rules to engage in any activity beyond that of 'first sale' - and the older cooperatives, for whom no such restriction applied. At Lorient, for example, the artisanal PO, PROMA, worked with a merchanting cooperative SCOMA, whose job it was to help the fishermen to get better prices. SCOMA competed with other merchants at the auctions and when there was a glut, bought at the minimum price and put the goods on the market for frozen products. In 1976, PROMA withdrew 305 tonnes for human consumption at a value of 430,000FF, but SCOMA, acting on behalf of the PO, withdrew fish worth 1.6MF and put it on the frozen market. Without this help, PROMA would either have run out of money or been obliged to double the contributions of its members.[31]

The absence of this kind of mechanism undermined the ability of the British POs to act effectively. In turn, the lack of perceived effectiveness intensified the 'free-rider' problem, with non-members free to sell at any price and thus able to depress the general level of prices. The result was continuous financial difficulties for the POs. Despite receiving £600,000 of the March 1980 aid package, the SFO soon ran out of cash again. In these circumstancs, as Fishing News pointed out, any fishermen was bound to ask himself what the point was in belonging to the organisation: it could not guarantee prices and no amount of money seemed able to make it viable.[32] The heads of the POs continued to complain bitterly about unofficial sales under the minimum price. And yet they could not point to clear material benefits which might persuade the 30-40% of fishermen estimated not to belong to any PO to surrender their individual interest in favour of a wider collective interest. Fishermen were no more eager to pay up for the POs than they had traditionally been willing to increase their payments to the WFA, when it appealed for an increase in its levy on landings.

The French fishermen were no more virtuous than the British ones and unofficial sales, 'ventes sauvages', did continue. Nevertheless, there was a greater feeling of the success of PO activity as a way of helping the fishermen's income. Again this was not because of any innately superior powers of organisation but because these organisations were supported by the state, both ideologically and financially. The proposal of one observer of the British fishing scene that "the UK POs would do well to consider the French POs and to model their activities on them"[33] was no easy one to bring about. Such a reform would require that a whole way of thinking about the relation of an industry to the state be transposed, as well as the institutions themselves.

4. Interests beyond national production

Up to this point we have assumed that self-help could be understood solely in terms of the conflict between the interests of producers as a group and as individuals and that the resolution of that conflict depended very much on the attitude of the state. However, it will be suggested in this section that the earlier arguments need to be broadened to include the relationship between the interests of producers, in general, and the interests of those who are not involved in production.

The liberal economic stance of the British state did not make it possible for fishermen to develop a common producer interest. What it did do was to encourage the industry to pursue much more individualist forms of self-help, which did not necessarily help to maintain national production. Such individualism was very much frowned upon in France. It was part of the protectionist French ethos that priority be given to national production over other interests and any attempts to subvert it were viewed with suspicion.

4.1 The structure of the two industries

Despite their different economic orientations, both Britain and France were relatively open to imports of fish. They were unable to avoid the surge of imports which followed the growth in the export potential of countries such as Canada and Norway in the late 1970s and early 1980s. The catching interest in both countries tried to limit this increase in imports but the only result was increased conflict with other interests, particularly fish merchants and processors, who saw their job as guaranteeing supply to consumers at the best price. Thus merchants were particularly upset by the catchers' attempts to push up prices, notably through the PO system. The fishermen, for their part, maintained that the prices they were getting through the auction system were not sufficient to allow them to continue operating. There were regular appeals for those further down the economic chain to show solidarity and to rally round the catchers. As the Chief Executive of the SFO put it: "It's time for British processors to stand up and be counted along with fishermen to get to grips with imports."[34] The buyers would retort that what they were paying reflected market conditions and not any conscious attempt to undermine national catching capacity. As one put it in the summer of 1980:

> "The message at the moment is that the consumer is not prepared to pay the high prices for fish we experienced in the 70s. Buyers can only pay what the market will accept and today that price is low.
> ... we cannot afford to subsidise fishermen with high auction prices."[35]

In France, the conflict was the same, though the attempts to escape the dilemma were rather more varied. While the merchants refused from time to time to attend the auctions, the fishermen tried to organise direct sales to the public, something the Leclerc supermarket chain assisted and encouraged. However, these similarities between the two countries' conceal the fact that non-producer interests were much more entwined with those of producers in Britain than they were in France. The situation was particularly obvious in the deep-sea industry, where there were strong links between processing interests and the British fishing companies. The existence of a large frozen fish market encouraged firms such as Unilever, Christian Salvesen and Bird's Eye to establish a direct or indirect interest in the catching sector, so as to guarantee supply for their downstream operations (cf p44 above). BUT the largest of the trawler companies (149 vessels in 1974), was a direct subsidiary of Associated Fisheries, which had interests in the Ross foods group; Bird's Eye, though it had no fleet of its own, needed as much as 40,000t. a year to maintain its leading position in the UK frozen food market and so negotiated suitable contracts with the deep-sea owners.[36]

The sheer scale of the operations was not matched in France, where even in Boulogne, the centre of the French processing industry, the involvement of catchers outside production remained relatively limited. There were negotiated deals with processors but they were necessarily on a smaller scale, given the smaller size of French trawler companies. Even the largest firm in Brittany, Jego Quere never had more than 20 ships, as we have seen (p 60 above). As for the processors, they made some moves into the catching sector, but it remained a minor operation compared with the British equivalent. Saupiquet, for example, the largest tinned fish producer in France, operated only three freezers, while Peche et Froid, a firm with turnover comparable to that of Associated Fisheries (400MF), owned only five vessels.[37]

The contrast between the two countries necessarily made the British deep-sea companies more ambivalent about the notion of producer self-help than their French counterparts. In the EAPO forum, the French complained that the PO of the British deep-sea industry, the Fishery Organisation Society Ltd, was allowing imports into Hull and Grimsby at prices below the EEC withdrawal price. However, the POs own lack of catching power and its close links with the world of the processors made such behaviour unsurprising: the idea of maintaining supply to traditional customers prevailed over any notion of keeping up national production.[38]

This dilution of the producer interest was evident in Britain in the inshore as well as the deep-sea industry. At the beginning of this study (p58 above), it was stated that the defining characteristic of the inshore industry is that the boat belongs to one or more member of the group who man it. Although in France the increasing cost of vessels have enouraged ownership within the framework of cooperatives, the statement is basically correct. In Britain, by contrast, the situation is more complicated.[39] In many English ports and some Scottish ones, ownership is shared with, or even assured by, independent agencies. The agencies are small land-based companies, to which the fishermen have turned to look after the practical problems that their profession entails, be it selling the catch, keeping accounts or preparing the ships for sea. In exchange, the agencies receive a percentage of the vessel's gross earnings. So widespread is the practice that Oliver claims that in the area he studied, "scarcely any vessels larger than the cobble class operate totally independently of the agencies."[40]

The agencies offer more than just a useful service. At a time when vessels are making little or no profit, they can act as secondary bankers and protect the skippers for whom they are working from going out of business. What is more, it is a job which offers only small profits and very high aggravation. Any hint of dishonesty drives the skipper immediately to one of the agency's competitors.

However, the fishermen themselves are locked into a relationship which offers them very little opportunity to influence their own economic environment. Unlike the situation in a producer organisation or cooperative, they have no say in the running of the agency, nor any direct financial stake. Hence the policy of the agency may conflict with the fishermen's own interests as producers without their being able to respond, other than by turning to another company. Oliver indicates for example, that agencies are certainly not opposed to dealing in imported fish to supplement their income:

> "One agent admitted that his commission from overland fish in 1979/80 had been £100; in the 1980/81 financial year it had been £18,000, and most of this had come from imported fish. He admitted that importing fish was not in the interests of the local industry but said that it was the only way that he could stay in business at the time. Of the two evils this he thought was the lesser; if he had gone down, so would the nine vessels in which he had interests, and their skippers and crews."[41]

The example of agencies within the British inshore system illustrates a further barrier to the development of self-help through the system of producer organisations. The British liberal tradition of competition between economic agents created conditions in which the clear delineation of a producer interest within the fishing industry could not occur in the way in which the French protectionist ethos permitted. The structure of both the deep-sea and the inshore sectors in Britain encouraged the link between catching and trading interests and thus weakened resistance to imports. Hence efforts at self-help followed a different course from those in France.

4.2 The behaviour of the two industries

In discussing the British fishing industry at the beginning of the 1970s, a French observer was particularly struck that foreign trawlers could land fish anywhere in the United Kingdom without any particular formalities.[42] Such a liberal regime was in marked contrast to arrangements in France. To accept imported fish meant having a special 'import card' from the authorities. Though this card was abolished - amidst much dispute - in 1975,[43] the protectionist idea that it embodied continued to play an important part in determining the kind of behaviour that was considered legitimate by the French industry. Similarly, the two economic traditions conditioned the fishermen's attitudes of who should be allowed to fish. In France, the norm was for French boats with a French crew to return their catch to a French port and any exception demanded an explanation. In Britain there was a much more relaxed view towards the criterion of nationality and the major question was whether the activity made economic sense to the company or individuals concerned.

During the late 1970s and early 1980s fishermen in both countries looked for alternative ways of disposing of their catch. Here was a different way of the industry helping itself to cope with difficult economic circumstances. However, the scope for such action was not the same for the two industries. It was very much more difficult for the French to pursue such alternatives than it was for their counterparts in Britain. Indeed British practices appalled the French industry, who saw them as illegitimate, if not illegal. It was far from eager to encourage such initiatives at home.

In France the recognition that disposal of catches abroad could be profitable was particularly marked amongst the deep-sea owners of Lorient in Brittany. The port is far from the main European distribution networks and - like the ports of Scotland - suffers from major variations in prices. Considerable amounts of fish have to be regularly withdrawn for conversion into meal. In 1978, Besnard, the major Lorient owner, started to organise sales abroad, particularly to Germany and Britain. At the time, haddock, for example, was fetching the equivalent of 4.50FF per kilo in Britain and 1.50FF in Brittany. However, there was immediate concern that such landings would threaten the employment of dockers and merchants in Brittany. FROM Brittany met and agreed that any such sales would have to get the approval of the FROM i.e the owner himself was not free to determine the scale of overseas landings.[44] Moreover, the state effectively gave its blessing to a restriction of this kind when the issue reappeared on the agenda. On a visit to Brittany in the autumn of 1983,

Lengagne, the Secretary of State for the Sea, made it clear that such landings were only admissible as long as the capacity to process the catch did not exist in Brittany.[45]

Besnard met similar kinds of difficulty when he tried to use Scottish ports as 'forward bases' for his trawlers so as to cut costs. It took considerable negotiation with the unions to satisfy the crews and an agreement was only worked out when he contracted to fly the crews home for 15 days after every 27 days they spent at sea. Similarly, the domestic processors had to be calmed by arranging to transport the fish by lorry back to Lorient for 'value to be added' to the catch. Despite the technical difficulty of keeping the catch fresh, this was the price to be paid for reducing the fuel bill of Besnard's vessels.[46]

The obligation of negotiating a code of practice <u>before</u> undertaking new initiatives of this kind was in marked contrast to what happened in the British context. In Britain rules of behaviour, if they were thought necessary at all, came <u>after</u> such initiatives were attempted. Thus British ships continued to land fish wherever they thought they could make most money and this practice did not stimulate any lengthy debate. As one fisherman commented, when asked about illegal landings of herring in Denmark:

> "I'm not saying it's on a grand scale as far as the British concerned, but they are involved. It's not possible to bring the fish to this country so they take them to the Continent for the good markets. Most men feel that if the continentals are being allowed to fish for herring then why shouldn't the British."[47]

A similar example was the landing of some of the major mackerel catch in nothern France. Though it was not illegal, the French response was one of considerable concern at the likely effect on prices in the market: for the Scottish seiner involved it was a question of where the captain thought he could make the most money.[48]

A much more important British initiative was the development of transhipment of mackerel onto Eastern European trawlers without the fish being landed at all. Until 1977, this fish was mainly caught by the factory ships from Eastern Europe. However, the extension of British fishery limits to 200 miles in 1977 and the failure of the Eastern Europeans to reach any agreement with the EEC on continued access led to their exclusion from Community waters, including those containing mackerel.

Thereafter the continuing needs of the Eastern bloc trawlers were catered for in a different way. Some of the Hull Humberside trawlers which were no longer able to fish for cod off Iceland joined some of the Scottish inshore boats and followed the mackerel each season from Ullapool to Falmouth. The fish they caught was far more than the domestic market could cope with and so they arranged to transfer the catch directly onto the factory ships. This proved to be a very profitable business with the catch rising dramatically from 326,000 tonnes in 1977 to 606,000 tonnes in 1980.[49]

The French industry saw this as an activity escaping any effective national control. FROM Brittany, for example, maintained that as much as two-thirds of the mackerel catch was not being declared and that as a result the quota was likely to be exceeded by 150 to 200,000t.[50] Certainly it was correct to say that the setting of rules to govern the activity followed rather than preceded its initiation. The organisation was highly decentralised and depended on a whole series of independent agents who negotiated prices with the Eastern Europeans. In some cases, fishermen themselves set up their own selling company and went out to attract further interest in the mackerel fishery: in 1979, for example, the Taits from Fraserburgh established a company, Falmouth Fish Selling, to persuade five Rumanian ships to come and then sold them over 1,000t. per week.[51]

It was not until February 1982, as the mackerel catch began to tail off, that a system of licensing was introduced for the factory ships involved and this fact alone illustrates how much more conducive the British environment was to this kind of self-help. The argument that counted was economic necessity, not any worries about where the fish was destined. As the Executive Director of one of the agencies pointed out, the vessels provided:

> "an outlet for about 70 per cent of the total British catch of mackerel and without them the severe crisis that the fishing crisis is going through would have ended in complete collapse for large sections of the industry two or three years ago."[52]

By contrast a much smaller deal with a Soviet firm to sell 3 to 6,000t. of herring from Boulogne caused considerable consternation. This was so despite the fact that as with the mackerel, there was no way of getting rid of the fish by normal export or internal consumption and the price was better than that for conversion into fishmeal (5FF per kilo rather than 1.75FF). The owner concerned was not free to carry through the deal as he wished but had to come to an arrangement with the rest of the local industry, giving an assurance that the processing of the fish would take place in Boulogne to protect local employment.[53] Once again it was a case of behaviour deviating from the norm of national production: for fish <u>not</u> to be landed and processed in a French port had to be justified in a way that was never considered necessary in Britain. There the place of landing needed no more justification than that it corresponded to the needs of the market.

The value of national production in the French context was not only buttressed by norms of behaviour as to the place of landing but also by particular expectations as to who should be involved in fishing. The 'code de travail maritime' laid down that the crews of French ships must be of French nationality. Although this rule was challenged by the ECJ in 1974, claiming that it contravened the Treaty article on the free movement of workers, it remained a powerful influence on French thinking.[54] In 1982 L'Helgouach, the CGT chairman of the Local Committee at Concarneau, complained that Spanish fishermen were getting the 'livret maritime professionel', entitling them to work as fishermen. The Minister, Le Pensec, defended what was happening but he made it clear that the problem was one of a shortage of candidates for the training courses provided under the auspices of AGEAM and that these were exceptional circumstances.[55] By contrast, it was seen as perfectly normal and something of a 'coup' for Marrs' of Fleetwood to get the expertise of some French sailors, using French gear, to help a British ship to fish in unfamiliar waters off the West of Scotland. There was great delight that the ship 'Irvana' returned with a catch valued at £48,872, a delight only matched by angry French reaction at the traitorous conduct of their compatriots in passing on their know-how to the English.[56]

However, the French raised a much more serious complaint about the registration of Spanish trawlers in Britain. This practice revealed to what extent other interests were intertwined with those of national fish production in Britain and helped to prevent a common producer front emerging. Though in French eyes it appeared to be a case of deliberate use of the law to increase British fishing capacity, it was in reality a reflection of the importance of a wider shipping interest which was only amended with difficulty to take account of the narrower fishing interest.

The practice consisted of Spanish trawlers being bought by companies, generally based in the Channel Islands, being reregistered at a British port and with a British name and then being sent to sea to fish with a British captain but a Spanish crew. It first gained serious attention in 1980 when fishermen in Devon and Cornwall started to express concern about the impact of the landings of these vessels, mainly registered and landing their catch in South West ports, on local fishermen. Despite reassurances from the companies concerned, the pressure for something to be done grew in 1981, with the Euro MP for the area, Harris, particularly active in pressing for change. Some amendments to the rules were introduced but the number of ships involved continued to increase. By the end of 1982, there were 60 ships operating in this way and it was not until 1984 that the first instances of deregistration took place, when the companies concerned were found to have only the most cursory link with Britain.[57]

These developments were of no small interest to the French because the ships fished off the Azores in an area where their own industry was active. They found it incredible that such practices were possible given French attitudes towards the nationality criterion. They therefore assumed that it was a manoeuvre designed to increase British negotiating strength in the EEC arena. As FROM Brittany put it:

> "The UK no longer has a fishing fleet of sufficient size (for the quotas it demands) and now openly appeals to ...Spanish mercenary vessels to which it fastens the British flag so as to try to succeed in fulfilling those quotas."[58]

However, this charge underestimated the importance of a pattern of rules which did not distinguish between fishing and other types of vessels. The responsibility for the registration of vessels belongs not to MAFF but to the Department of Trade (DoT) and its own priorities are determined by the Merchant Shipping Order of 1927. This piece of legislation had specifically wanted to encourage an increase in the number of ships on the vessel register, and had therefore specified only that the skipper and master needed to be British. The DoT remained reluctant to tighten the regulations on the grounds that it did not wish to discourage investment and wanted to make foreign shareholding possible in the ship industry.[59] In early 1981, it did agree to impose the condition that the companies concerned should be able to show that their principal place of business was in Britain but it was unwilling to go any further to modify legislation which affected all shipping and not just fishing vessels.

What reinforced an unwillingness to discriminate on the basis of nationality was the feeling that the ships were not necessarily a threat to domestic producers. As an agent for one of the firms put it:

> "We are doing the local fishermen no harm. We do not fish anywhere near their grounds. I am providing good business to Penzance and providing local employment."[60]

Local fishermen certainly did complain and increasingly so, but their argument was not based on the criterion of nationality. Rather they felt that the argument of the firms was disingenuous. Although the fish was not sold on the open market, it necessarily depressed business and prices at the auction where locally-caught fish was sold, especially when so many ships were involved. However, when it came to discussion at the European level within EAPO, it was the French rather than the British who pressed hardest to have a rule for the registration of boats so that 75% of the capital and of the crew must be of EEC origin.[61] British opposition to the scheme was more muted because it was clear that it did benefit some in the industry. Former Humberside captains were helping themselves by signing up on the Spanish vessels and it was not so clear that they should not be allowed to. Certainly the nationality of the vessels was not the central issue as it was in French arguments.

What has emerged in this section is that the French state stood behind a pattern of economic activity which the fishermen themselves maintained. It did not require the state to express formal opposition to more liberal practices. The industry internalised resistance to them, channelling its self-help efforts through the producer organisations. In Britain, by contrast the idea of national production was only one element and not a strong one at that. Producer organisations were developed alongside agencies and companies, all seeking to help themselves in the face of economic change. This itself was possible because the state never took a very firm view as to how the industry should operate: that was for the industry itself to decide. The very diversity of the decisions that were taken by the operators only serves to underline the power of liberal economic doctrine in Britain. The attempt in France to reduce or control that diversity, a reduction undertaken by state and producers alike is what justifies the claim that 'dirigisme' remains a guiding principle of action there.

5. Conclusion

This chapter has come to two main conclusions. First, it has broadly confirmed our expectations as to the behaviour of the industry. There was a realisation in both countries of the value of self-help. However, the shape of that self-help varied enormously in the two countries. In France stress was laid on the potential of the EEC POs as a way of jointly organising the market; in Britain the activities of the industry were more individualistic and less coordinated. The French protectionist tradition made it considerably easier for the POs to organise themselves and to discourage production outside an organised framework. In Britain, by contrast, the liberal tradition meant that control over the activities of separate economic agents was not encouraged and as a result, the concept of a fish producer interest, clearly distinguishable from other interests, made much more limited progress.

Secondly, this chapter has emphasized once more the importance of the idea of the state in this study. It has shown that societal self-regulation cannot be disentangled from the nature of the state environment within which it takes place. The French and British states heavily determined the shape and behaviour of self-help structures. In France particular stress was laid on the idea of fishermen as a group, thus reducing the scope for individual initiative; in Britain the concentration was not on fishermen as a collective but as a set of distinct individuals, with different problems and priorities. Once again the British state acted as an arbitrator between separate interests in their efforts to help themselves rather than seeking to set the terms of such action in the 'transcendental' style adopted in France.

9 State and society in the fishing industry: an overview

1. Introduction

The final chapter will review the study as a whole and examine two aspects of the relationship between state and society in the fishing industry. First, it will look at the pattern of behaviour of government and industry in the two countries and suggest that it displayed an important degree of continuity. In this context, it will stress the importance of different historical traditions, the power of the national arena as the focus of attention and the links between the four perspectives used. Second, it will consider the issue of success in devising a response to change in the industry. It will point to three dilemmas over conflicting priorities that received different answers on the two sides of the Channel: the choice between production and consumption, the balance between public and private power and the clash between the exploitation and conservation of a common resource.

2. The pattern of behaviour

Although one of the aims of the study was to examine the impact of a period of major change upon the behaviour of an interest in society, one of the results has been to stress the degree of continuity in the pattern of relations between that interest and the state. By considering what we might have expected to happen to that pattern, we have certainly found changes of emphasis. Actual or threatened direct action, for example, became more widespread than it had ever been before. However, what is more conspicuous is that the differences between the two countries remained as marked at the end of the period as they were at the beginning. By itself, an economic crisis is not sufficient therefore to explain why those affected by it respond in the way that they do.

2.1 Historical tradition

What the continuity in relations shows is that a historically-established pattern was not easily overturned, as some expected or desired. At the beginning of the 1970s Mordrel, for example, argued that France's increasing openness to the world economy would combine with the growing internationalisation of the market in fisheries to break down the interprofessional system established in 1945 and the protectionist outlook that it embodied.[1] He was mistaken. The resistance of the French industry to the encroachment of liberal economic pressures remained remarkably strong and a certain weakening of the CCPM structure was made good by the development of new corporatist structures in the form of FIOM and its privileged relations with the French producer organisations. True the increase in the volume of imports was not stopped as a result. But opposition to the 'Europe des marchands', i.e a traders' Europe, remained a significant principle of action throughout an industry still united - employers and unions, inshore and deep-sea - around the promotion of production. Their joint opposition in 1980 to the liberal ideas of the Barre government, and their continuing commitment to the protection afforded by a fuel subsidy, afford eloquent testimony to the maintenance of a system whose demise Mordrel had predicted nearly a decade previously.

Similarly, the repeated refrain in Britain that the industry should become more united and overcome its notorious level of fragmentation met only partial success. New federations were founded - notably the NFFO and the SFF - which did succeed in generating a degree of unity in the inshore industry but the common denominator between their members remained at a rather low level. As we have seen (p 207 above) their leaders flew in strength to Brussels and Luxembourg every time the Council of Ministers met to keep up the pressure for a favourable settlement. However, the presence of so many was not just a sign of improved organisation; it also revealed the weight of the fear that some would be favoured more than others when agreement came. In this sense, the EEC provided a convenient scapegoat for a very long time, and hid the extent of the differences that continued to separate British fishermen. The British government might deplore those differences but it did very little to limit their fullest articulation.

Both examples show the strength of the separate historical traditions that prevailed in the two countries, traditions which strongly influenced perceptions of those faced with the need to respond to change. At the beginning of 1975, for example, both countries suffered from a large volume of imports of fish and a consequent drop in prices. France's response, notably in the creation of FIOM and the establishment of a substantial fuel subsidy, reflected a protectionist past which Britain did not have. Hence the British government did not perceive an equivalent response as in any way suitable and persisted in a more 'liberal' form of intervention. Similarly, in 1980 the French government was blind to the possibility of cajoling the fishermen into accepting limited concessions in return for giving up their direct action. The 'limited authoritarianism' of the French state excluded the kind of flexible response which marked the behaviour of British ministers in 1975 and 1980. The problems for the two governments were similar but their different economic and political pasts made it hard for them to devise equivalent solutions.

The importance of the distinct traditions of the two countries becomes still clearer if we consider the concepts of 'corporatism' and 'overload'. It was suggested earlier (p 8) that the literature linked to these concepts tends to underestimate the differences between Western European states. This view is confirmed by the evidence for continuity which has emerged here.

The 'corporatist' thesis that governments are drawing interests more closely into the business of government because of the difficulty of managing the modern economy, requires an awareness of the evidence uncovered in this study. In France, the pattern of relations in the fishing industry already had a corporatist imprint long before the 1970s. The 1945 institutions, a suitable variant on what the Vichy regime had set up, implicated the industry in the state's decisions and offered the state ample opportunity to influence the industry's attitudes. In the period of this study, the closeness of the links continued to limit outside involvement in the fishing sector. In Britain, by contrast, representative channels, provided by MPs and Parliament, not only remained in use but were still more intensively exploited by an industry that saw no better way of making its position known and accepted. The state remained very conscious of the importance of these channels and did not try to short-circuit them, despite the difficulty of resolving the problems of the industry. In both countries, certain features in the pattern of relations were accentuated but the basic shape did not change.

According to the 'overload' thesis, Western governments increasingly lack the means to cope with the number of economic demands that the political base of their authority invites. The lesson of the study in this case is that such a general thesis needs to accommodate the relative tolerance of different countries to increased demands. Certainly in 1980 the French mechanisms of consultation were 'overloaded' and unable to cope with a crisis. But that was because the government concerned was attempting to reduce a high level of expectations as to the state's obligations towards the industry. The goverment that followed was able to revert to the old pattern and thereby satisfy the industry. The establishment of a Ministry of the Sea was seen as the expression of a caring form of 'tutelle', which for the industry stretched back to the 18th century and the age of Colbert. In the British case, not only was the level of expectation traditionally lower but there was a range of channels through which complaints could be filtered. The strength of the various representative leaders of the industry served to reduce the danger of the government being overtaken by the industry's demands. Moreover, ministers, in particular Walker, were conscious of the importance of this channel and responded to the industry's promptings in finding sums of aid at the appropriate moment. That the demands of the industry came at a time when the general economic situation was bad cannot be denied; that the capacity of government to respond to them was so reduced as to merit the term 'overload' is much less clear.

In this way the historical context of this study has underlined the value of an intermediate level of analysis which goes beyond day-by-day detail but does not draw broader conclusions as to the changing shape of societies in Western Europe. Such changes may be taking place, but the evidence here suggests that there are also important pressures for continuity which derive from the character of the relations between state and society.

2.2 The national arena

A second conclusion of this study, which also underlines the importance of continuity, is that the national arena remained the central focus of action on the part of the fishing interest. This may seem paradoxical when it was argued in Chapter Three that the entry of the issue into the EEC framework ended its treatment as a sectoral concern and heralded a new era of institutional interdependence. It may also appear to be contradicted by the ability of the ten member states to reach agreement in January 1983 on the shape of a CFP, designed to govern relations in fisheries until the next century. Leigh, for instance, has argued that "the CFP provides an example of tangible integration achieved by the indirect functional route."[2]

However, the fact that national administrations were put under pressure by the Community institutions and their Treaty obligations to reach an accord, should not be exaggerated in importance. As Chapter Three showed, agreement did not mean that national perceptions of the problems to be resolved did not remain crucially distinct. At the beginning of the 1980s France was still concerned to ensure protection of its national market through the EEC, as it had been ten years previously, while Britain remained committed to the importance of maximising limits and quotas and ensuring that conservation measures were as effective as possible. The final agreement had to contain elements that satisfied both views (as well as those of the other member states). Hence the idea to have Community fisheries inspectors was a British one just as much as the strengthening of discipline in producer organisations was French-inspired. Common policy did not correspond to convergent perceptions.

The relative weakness of the Community arena was underlined by the behaviour of governments domestically. They were certainly not eager to find themselves arraigned before the ECJ in Luxembourg, but equally they were prepared to act unilaterally if they felt such action to be necessary. In Britain the argument that the government was unfairly subsidising the industry was met by stressing the temporary nature of the aids. But on issues perceived to be more crucial, such as conservation, neither Labour nor Conservative ministers hesitated to take unilateral measures contrary to the Hague agreement. The French, for their part, were happy to abide, at least in principle, by conservation measures such as the herring ban, but made no excuses for the fuel aid that they accorded to their fishermen. Le Theule, the French minister, might quote EEC policy as a justification for not increasing that aid in the summer of 1980 but the actions of those who preceded and came after him indicated that he was in reality seeking to deflect criticism from the government and using the Community as an excuse. What determined national policy were domestic, economic and political considerations, not a supranational authority.

The fishermen, for their part, were equally reluctant to transfer their loyalties outside the national arena. When they attended Council meetings, it was to back their national minister or at least to prevent him backing down, not to support any Commission compromise. When they went to 'Europêche' or EAPO, it was as much to gain a wider airing for their own domestic complaints - be it the Spanish vessel registration scheme or the illegal landing of herring at Boulogne - as to develop common

European policies. They were well aware that the differences between the representatives of the industry were likely to be greater at the European than the domestic level. As French inshore fishermen commented as long ago as 1967:

> "Do we find in Brussels and elsewhere people who want to listen to us? Instead of owners and above all, active fishermen, more often than not we discover that the people we speak with are only representatives of the traders (i.e. Unilever, Findus, etc) who have no interest in the fate of producers and with good reason."[3]

Nearly 15 years later, the instincts of French producers were no different when the head of the cooperative organisation (COCMM) remarked: "If we do not act, in a few years the French fishing industry will be a marginal activity integrated into the workings of multinational companies."[4] In such circumstances, there was little incentive to establish any kind of clientelist relations with EEC institutions to replace their activity in the national arena.[5]

The argument presented here does not seek to deny that the character of the issue changed, that there was a significant 'blurring of the boundaries' between foreign and domestic policy within the EEC framework. Nor does it attempt to suggest that national solutions were necessarily a viable alternative to the CFP, or that the levels of integration may not be increasing and increasable. What it does do is to challenge the easy assumption that the move towards agreement in 1983 was the product of some linear process of integration, which can be separated from the domestic relationship between state and society in the fishing industry.[6]

2.3 The link between the perspectives

The final conclusion in this section returns to the point made at the outset (Chapter One, p29) that the four perspectives used in this study should not be seen as separate entities without any relationship one to the other. Rather they form an inter-related set of ideas which together define an overall logic of action governing the relationship between the government and the fishing interest in the two countries. Here too we can see why that relationship was not fundamentally affected by the changes that occurred in the issue area.

The two perspectives which stressed the economic behaviour of state and industry were presented in Chapters 5 and 8. The first of these assumed a 'top-down' view, stressing the idea of the state's intervention in the affairs of the industry, while the second took a 'bottom-up' view, considering how the industry organised to help itself in the face of economic change. It emerged from the discussion that neither state intervention nor the industry's self-help could be understood on their own without reference to the other. The industry's expectations as to the state's intervention were as important as the state's attitude towards the industry's self-help. Moreover, differences between these expectations and attitudes explained why behaviour varied so much between the two countries. The French industry had traditionally looked to the state for economic support and this pattern remained as clearly visible as the reluctance of the British industry to see such support as anything but a necessary evil. Thus the 'barons' of the British deep-sea sector were at one with the individualists of the inshore sector in pressing for temporary injections of cash to keep the creditors at bay, just as the UAP could agree with the French 'artisans' on the need for maintaining or increasing the subsidy to maintain their productive potential. Similarly, the supervisory nature of the state's economic role in France encouraged it to provide strong financial and institutional backing for the industry, as it sought to organise itself more effectively, while the British state's aversion to anything beyond limited intervention meant that there was considerably more scope for individual skippers and companies to help themselves but considerably less for fishermen to establish a collective economic identity as producers. The mutual recrimination between the two industries is hardly surprising when these contrasting attitudes towards intervention and self-help are considered.

The political response of state and industry was the centre of attention in Chapters Six and Seven. The former looked at the conventional channels for exerting influence on government, considering how the grievances of the industry were mediated and represented, while the latter examined the extent to which those grievances spilled over into illegitimate behaviour with direct action serving to underline forthright opposition to the state.

As in the case of the 'economic' chapters, a link was found between the two cases with legitimate action mirroring the unconventional and vice versa. The centrifugal pressures at work in the British industry generated a group of powerful independent regional representatives, who could ensure the widest ventilation of grievances through the parliamentary arena. At the same time, their independence helped to control demands for direct action, while the openness of the conventional channels served to reduce the potency of such demands. By contrast, the French state's capacity to incorporate the industry's representatives left them with little scope or desire to exert their independence from the authorities and meant that they enjoyed no integral connection with other mediatory bodies, such as the political parties. Rather they were passed by as the discontent of the industry built up, with no mechanisms equivalent to those in Britain available to dissipate it. The French state, for its part, was obliged either to surrender to the pressure or to crush it and could not use the combination of limited concessions and the leadership's consent to them which proved so successful in the British context in calming dissent.

However, the links between the perspectives go further in that the pattern of behaviour in the political arena should not be disassociated from that in the economic area. Thus the liberal pattern of free competition between economic agents that was central to thinking in Britain, was matched by the political behaviour of the industry's representative institutions. Both were premised on the need for a decentralised system of power, which took account of the variety of the interests involved. However much greater unity might be praised, however loud the calls for greater protection, ideas of new arrangements which would seriously limit the economic and political choices of the industry made little headway: hence the continuing disputes between regions and the lack of bite in bodies like UKAFPO. By contrast, the protection enjoyed by the French industry cannot be separated from its corporatist form of political organisation, where the differences between the various sections of the industry were muted in the interests of centralised forms of control. Acceptance of the need for limits on choice meant that the pattern of representation changed no more radically than the pattern of economic behaviour: the CCPM stayed intact just as Besnard's landings in Scotland continued to be seen as an aberration. Thus France remained very much a 'state-led society', in both a political and economic sense, just as Britain showed that she was still a 'society-led state'.

To make these points is to stress again the claim that none of the perspectives should be seen as representing a norm, from which the others diverge. If there is such a norm, then it is the logic of action in the two countries which emerges when all four perspectives are considered together. Within each country the balance of behaviour may change over time but what differentiates them most clearly is the overall shape of the relationship between state and society. That shape cannot be derived from one perspective alone.

3. The issue of success

The second major issue of the study that will now be reviewed is the success achieved in devising a response to change in the fishing industry. Even though such a question involves an evaluation which cannot be decided by analysis, it is the author's view that any analyst has a responsibility to make the foundations of such an evaluation clear. In fact, the possibility of comparison between two countries with such contrasting political and economic arrangements makes it easier to uncover those foundations than would be possible if only one of the countries was being examined. They will be seen to have stressed different ways of treating the fisheries interest, and no one can hope to judge success without recognising the incompatibility of these competing emphases.

3.1 Production and consumption

It has been one of the themes of this study that though it has treated the fishing industry for much of the time as a unitary entity, in reality the industry is composed of a whole series of different parts which by no means always share the same aims and objectives. Once this diversity is accepted, the question of success has to be modified so that one can ask: success for whom and through whose eyes? Even inside the British catching industry, there were many differences of opinion as to the extent of the limits and quotas which were necessary to guarantee a secure future, depending on the region concerned and the kind of fishing involved. It is thus futile to try to ground a judgement of success solely on the perception of those involved: there is no way of aggregating a mass of conflicting opinion of varying intensities.

What needs to be done is to consider the purposes that fishing can be perceived to perform and the way that the state arbitrated between them. There are two distinctive ways in which the activities of catchers can be viewed: they can be seen either to contribute to the market supply of a product or to maintain a market through their production. In the former case, production depends upon, and must be conditioned by, the requirements of the market; in the latter case, the market is the result of, and has to be adapted to, production. This distinction assumed a much greater importance as fisheries was more firmly integrated into the international economy during the 1970s. The trade in fish products increased markedly and national markets became ever more vulnerable to economic developments on the other side of the globe. The state's basic predilection for production or consumption was put to the test and highlighted by a major shift in the economic conditions facing the industry.

In the first chapter (p 7), the question was asked: were the two industries doomed to decline? Even on the most optimistic assumptions of what Britain could have done outside the EEC, the answer without question seems to be positive for both countries. The increase in pressure on stocks, the spread of the EEZ doctrine and the rises in the price of fuel make it hard, if not impossible, to imagine that the two fishing industries could have continued to enjoy the relative prosperity of the early 1970s.

Nevertheless, it is clear that the position was not entirely without hope: there were favourable circumstances which could be turned to advantage. The French tuna fleet, for example, was able to emerge into the 1980s in as good a shape as it had been in the early 1970s. It offered a marked contrast to the rest of the deep-sea industry in the two countries. At the same time, those sections of the inshore industry which caught valuable species that were not subject to the competition of frozen fish could expect to continue to maintain high incomes and an important place in the fish market.

However, the question was also asked: could the industries have been better protected? The answer to this question obliges us to consider the extent to which the two states were prepared to go to maintain the shape of their fishing industries in the face of change. Yes, British governmnts could have provided better protection for their industry, not by negotiating harder in Brussels or sending more gunboats to Iceland but by changing the balance of their policy between the philosophies of production and consumption. As the discussion in Chapter Five indicated, there were opportunities - not necessarily cheap or easy ones - for maintaining at least some of the productive potential of the fishing fleet without pushing other EEC countries out of the NEAFC area. But these were perceived as unrealistic or non-existent because there was no shortage of supply to the market. Over the period of this study, the overall catch level was allowed to fall from over 1,100,000t. to under 800,000t. but the level of supply to the market remained roughly constant (cf p45 above). There might be fewer fishmongers' shops but there was no drop in the level of landings to keep consumption up: the merchants and processors stayed in business, even with the virtual disappearance of the deep-sea fleet.

As for the French state, it was clearly more heavily committed to maintaining the basic shape of the industry and setting production before consumption. Despite an increase in imports, the level of national production did not change very much: in 1982 landings were around 700,000t. just as they had been in the middle of the previous decade (cf p 47 above). This was not achieved without paying a price. By 1982 there were just over 20,000 'marins', 20% less than the number six years previously, and the number of vessels had declined markedly from 13,268 in 1974 to 10,873 in 1981, figures reflecting a greater decline than in Britain (cf p 61 above). However, the effect of the drop in men and boats was to increase productivity and thus to improve the chances of maintaining national production in the face of competition from abroad.

The French productivist ethic was reinforced by the contrast of attitudes between the inshore and deep-sea industries. The sharing of risks between capital and labour on board the vessel, the possibility of the crew member himself becoming an owner, and a system where remuneration is directly related to the value of the catch, combined to maintain an inshore sector whose members were prepared to undergo considerable hardship, long hours and few holidays in order to make as large a catch as possible. The deep sea crews, on the other hand, had none of the same incentives and as many, if not more, of the disadvantages. Their relative decline therefore bolstered the importance of the idea of production as an important value in the French industry as a whole.

The motivation of individuals in the British inshore sector was similar, if somewhat diluted by the existence of the agency system in some parts of the country. However, the disappearance of the deep-sea sector did not result in an increase in the productivist ethic. Though the Left in Britain, for example, castigated the conditions in the company sector as "antiquated, vicious and corrupt and lethal",[7] the solution was seen in terms of decasualisation of the deep-sea vessels rather than encouragement of the values of the inshore sector. There was no pleasure at the passing of a brutal mode of production, but rather annoyance that its chances of survival seemed to have been undermined by a deadly combination of rapacious Icelanders and continental Europeans.

Thus the success of the response to change cannot therefore be divorced from the purposes seen as central to that industry. Two examples from the previous chapter can clarify the point. To applaud the relative failure of efforts by South West fishermen to end the system allowing Spanish vessel registration is in part to defend the liberal, non-productivist philosophy that permitted it; to feel that British fishermen should follow the example of ANOP is to value the idea that the state should foster the collective interests of producers through close cooperation with them. The two traditions are necessarily in contradiction with each other and any judgement of success depends on the relative value placed upon them by the observer. To the author, it seems that if value is placed on preserving a particular pattern of production within the economy, then in a period of economic contraction, the French tradition is seen to have been better able to achieve that end. If, however, the important thing is to ensure that consumers get a product at the cheapest price, then there is no reason to condemn the more liberal British arrangements.

3.2 Public and private power

The second component of an evaluation of success derives from the way in which power is exercised and the balance that is struck between public and private objectives. The achievement or non-achievement of private goals cannot be separated from the scope that is afforded to the state to achieve public ends. The greater the strength of the collective principle, the less the opportunity for individual adaptation to economic change. Once again there is a conflict which can only be uncovered and not be resolved by analysis.

In crude terms, it seems obvious that the British fishing industry failed to get what it wanted from the government as a basis for its future operations. The EEC deal that was reached was a long way from what it had wanted. Its demands for a 50 and then a 12 mile exclusive limit were not met: the negotiations produced a 6 mile exclusive limit with a degree of foreign access to the 6 to 12 mile belt. Its call for a fair share of a 60% stock contribution, equal to at least 45% of the fish available, also made only limited headway: an initial offer of 31% could not be pushed higher than 38% of the seven main species.[8] However, the government could fairly reply that the deal was a good one. The increase in the percentage of the total catch available to Britain was only made possible by obliging other countries to accept a cut in the quota available to them, and the 'equal access' principle had been substantially breached by limiting fishing in the 6 to 12 mile area to countries with historic rights and in a very sensitive area such as around the Shetlands, to a restricted number of licensed vessels.

We can argue about the relative strength of the two sets of claims but it is essentially not a very useful exercise. No doubt there was a degree of unreality in the initial position of the UK industry, no doubt the nature of the EEC limited what was seen as politically possible by the British government: both were the prisoners of their own perceptions. What is more interesting is the way in which the movement towards a settlement occurred. There was no attempt on the part of the state to define what the industry should accept rather it was a question of testing the ground to find what would be acceptable to the various interests involved. The process epitomised the activity of what Shonfield has called "a wheeling and dealing type of public authority constantly seeking out allies, probing and manoeuvring for the active consensus".[9] It was a good example of a state with an 'instrumental' view of the interests in society, seeking to arbitrate between them but not seeking to aggregate them into a wider collective vision.

The advantages of such an arrangement are not far to seek. First of all, the move towards a settlement could be brought about with the maximum degree of consensus. The leaders of the industry were given the opportunity to ventilate their grievances as much and as often as they wished but could then be persuaded by an effective minister into accepting the hitherto unacceptable and defending their own 'realism' in front of their sceptical members. Secondly, the members of the industry were left free to decide how to respond to the gradually emerging shape of the settlement. The members of the deep-sea industry, in particular, did not stand idly by as the number of active vessels declined. The crewmen departed in large numbers to the offshore oil industry, the captains sought jobs elsewhere in the industry, some gaining posts on Spanish-registered vessels, while the owners moved their money into the inshore-sector or away from fishing altogether into areas such as agriculture and building.

The consequences of such a system were not necessarily favourable. The passing of the deep-sea sector undermined the roots of local communities and provoked a degree of unemployment amongst men less well able to adapt to change than their well-heeled employers. While this was happening, successive governments gave the impression of standing idly by, not taking the interests of the industry seriously enough into account. This seems, however, to be a consequence that is hard to avoid unless one is prepared to challenge the value of a system based on the principles of political consensus and economic freedom of choice. Whatever their views on the EEC settlement, few in the industry seemed to want to mount such a challenge.

Although the French system of government does not deny those same principles, there is little doubt that the fishing case illustrates its capacity to give them a lower level of priority. The importance of the state in channelling political and economic demands created a different balance between public and private power in France. The fact that the state had always exercised an important level of 'tutelle' or supervision over the French fishing industry meant complaints against it were set in terms of an active rather than passive neglect. It was not that nothing was being done but that the authorities were proposing to threaten openly the interests of the industry, whether by reducing its size in the interests of modernisation or by making it face up to the 'real' costs of energy. As we have seen these efforts, prompted in particular by the Barre government from 1976 to 1981, provoked a considerable degree of disaffection, which exploded in the summer of 1980. The notion of consensus evaporated when the whole industry perceived a threat from the state.

Such a perception was, however, only possible where the state's supervision had generally been seen in a positive light. There was an almost universal confidence in the institutions established in 1945, which was reinforced as FIOM came to assume a larger role in support for experimental voyages and producer organisations after 1976. It was acknowledged that the powerful position of the state within such bodies helped to reduce internal rivalries within the industry and to prevent the much more uneven pattern of development seen in Britain from emerging.[10] The institutional incorporation of the sector's representatives was matched by a degree of economic protection and direction, without parallel in the British context. The economic freedom of choice of operators was set within a much tighter set of parameters, where the state not only intervened financially but was also eager that its intervention have a particular effect on the activity of fishermen: hence the choice of a fuel subsidy. The French state wanted not only to wind up the clockwork mouse but also to guide where it went.

The problem was that such a system developed an important level of rigidity. The CCPM strucure, for example, seemed ideally suited to guarantee the maximum level of consultation with the minimum level of opposition. But once grievances built up, the complex hierarchy of committees proved much less flexible than the more informal, _ad hoc_ arrangements found in Britain. Similarly, there was state-supported resistance to moves, like that of the owner in Lorient to base his operations in Scotland, and yet it was hard to see how such a port could survive in the long run unless such initiatives were both permitted and encouraged. The desire to offer protection to the industry was in conflict with a desire to allow operators to arrange themselves in the way they thought best.

Once again it is not enough to praise one country's practices and to chide the other's. Both involved a trade-off, where judging success depends upon the value one places on the importance of private and public choices. If one prefers economic allocation of resources through private channels linked to political change governed by consensus and consent, then Britain's response to change in the fishing industry was not unsuccessful; if, on the other hand, one considers that public influence over choice of economic goals is important, even if there is the possibility of strong resistance to that influence, then French fishing policy can hardly be adjudged a failure.

3.3 Exploitation and conservation

The final part of this evaluation of success in the response to change is linked to the idea of timescale, the period over which the judgement is made. Should one's perspective be short-term or long-term? The question is a critical one because of the nature of the resource concerned and the behaviour of those involved in pursuing it. It was one of the clear lessons of the 1970s that fish, although renewable, are not inexhaustible: once a stock has been exploited beyond a certain point, it will be impossible for it to recover and it will be threatened with disappearance. There is therefore a clear collective interest for the industry to avoid overfishing but it is one that conflicts with the individual interests of its members. The 1970s showed that no fisherman has an incentive to cut back his own activity; on the contrary, it is better for him to have a larger boat, which will increase his earnings and reduce his tax liability. However, though the effect is likely to be beneficial to him in the short-term, the long-term impact is to weaken the chance of continued fishing for all. What is the value of increased catch levels by themselves, if they are followed by an irreversible decline?

The dilemma is one that has exercised governments greatly during the period of this study and it is a tribute to the intractable nature of the problem that the EEC agreement of January 1983 was not able to address it satisfactorily. No government was enthusiastic about the idea of withdrawing vessels permanently from their fishing fleet and the result was that:

> "out of a total 250 million ECUs allocated under the CFP agreement to adjustment of capacity, redeployment of capacity, and restructuring, modernising and developing the fishing industry and aquaculture, only 32 million ECUs (12.8%) are allocated to the 'permanent withdrawal' of vessels over 12 metres in length."[11]

In other words, there was plenty of money allocated for new development but very little to cope with the acknowledged overcapacity of the Community fishing fleet.

The reasons for this reluctance are not far to seek but differ in emphasis between the two countries in the study. In the French context, the difficulty is the commitment of both state and industry to the idea of production. Whenever there was a move to limit catches, the argument against it from the industry was that it failed to take account of the 'socio-economic environment', in other words, it was contrary to support for the activity of fishermen as producers. However well the state authorities might recognise the scientific case for effort limitation, they could not deny that they were themselves strongly in favour of promoting national production in the face of increasing imports.

In the British case, there was not this same ambivalence in that market supply was seen as more important than production *per se*. Hence within Britain there was strong support in the WFA and the ministries for tight control on entry into a fishery through a system of licences. The problem with this proposal was its assumption that the Community countries could agree on the respective size of their fishing fleets. This was not the case. No one would accept a totally-controlled fishery system if it meant that there was no room for a large fraction of their fleet. Yet without such control, no state, including Britain, would be willing to introduce licences and a limit on vessel construction for itself alone. As the DAFS Fishery Secretary put it:

> "As long as the majority of fish stocks are in nobody's ownership, it is very difficult and probably indefensible for a given Government to discourage its own fishermen from building up their fleets, to make the best use of available stocks."[12]

The grounds might be different but the basic resistance to effective limits on exploitation was the same in both countries.

One way out of this impasse is to argue that real success depends on moving from centrally-imposed solutions to those devised by the coastal community itself. As Stiles puts it in his discussion of Newfoundland:

> "... a decentralised and fisherman-oriented management regime is infinitely preferable to one which filters down from 'above', that is from scientists and bureaucrats living far away from the realities of fishing as a form of livelihood."[13]

Such solutions seem all the more attractive when it is recognised that a regime of this kind has proved possible even between fishermen of different nationalities in the EEC. There was a considerable degree of conflict between the trawlermen at Port-en-Bessin in Normandy and the pot fishermen of Devon with their static gear, who were active in the same area in the Western Channel. Despite their differences, in May 1981, they were able to make a formal agreement, dividing up the fishing zones, by area and by time of year and ten months later, it had worked well enough for them to be able to renew it.[14]

Unfortunately, the attraction of this approach can only be a superficial one. This is not to say that the agreements were not valuable, that the two groups were not able to look beyond their own individual interests to a wider common interest. But the problem was susceptible to a decentralised form of management because it was very limited in its extent. There were less than

100 vessels involved, the resource implications of the restraint imposed were minimal and it was a bilateral and not a multilateral difficulty. Once the number of vessels is increased, once the costs involved for the participants rise and once it ceases to be a question of a discrete issue between two parties, the possibility of escaping the collective goods dilemma in the fisheries arena begins to disappear very rapidly.

In reality, there is no reason to suppose that individual nation states will necessarily accept limits upon their domestic activities in the interests of international collaboration, simply because they are shown good reasons why they should do so. It is not just that the nature of collective goods can vary, as Ruggie indicates[15] but rather that the basis of their action depends upon the nature of the domestic relationship between state and society. On the evidence of this study, neither a state-led society nor a society-led state appears to offer a very favourable environment for the resolution of the balance between exploitation and conservation of the fisheries resource within the NEAFC area.

Yet if such a resolution cannot be found, then there is every likelihood that the fleets of Western Europe, including those of Britain and France, will be drawn into a depressing downward spiral of declining catches and shrinking fleets. Then the very concept of success as applied to the period between 1975 and 1983 may become irrelevant in the face of an empty sea.

APPENDICES

Appendix I: List of references

Two points of presentation should be noted here. Firstly, all quotes that come from French sources have been translated by the author. The translation has been put in the text; the original can be found under the relevant reference. Secondly, the British civil servants interviewed asked not to be linked to particular pieces of information. When such information is used, the relevant reference is worded 'private interview'.

CHAPTER ONE

1. Quoted by SCOTT, I. (1980) 'The importance of the North Sea fishery resources' in WATT, D.C. (ed.) The North Sea: A New International Regime?, Greenwich Forum V, Records of an International Conference at the Royal Naval College, Greenwich, 2, 3 and 4 May, 1979, p 139.

2. This process of 'nationalisation of the seas' is fully discussed and documented in LA DOCUMENTATION FRANÇAISE (1978) 'Les Etats et la Mer - le nationalisme maritime', in Notes et Etudes Documentaires, Nos. 4451-4452.

3. BROWN, E.D. 'Maritime Zones: a Survey of Claims' in CHURCHILL, R. SIMMONDS, K.R. and WELCH, J. (1973) New Directions in the Law of the Sea, New York, Oceana Publications, Vol.III, p 161.

4. LA DOCUMENTATION FRANÇAISE (1978) op.cit. p 19.

5. The statistics to support the points made in this paragraph are drawn from the British Sea Fishery Statistical Tables and the French Statistique de la Marine Marchande. Other information comes from the papersof the industry in the two countries Fishing News and Le Marin.

6. The best account of the development of the CFP up to 1983 is to be found in LEIGH, M. (1983) European Integration and the Common Fisheries Policy, Croom Helm, London. See also FARNELL, J. and ELLES, J. (1984) In Search of a Common Fisheries Policy, Aldershot, Gower.

7. A good example of the genre which is orientated towards particular case studies is KIMBER, R. and RICHARDSON, J.J. (eds) (1974) Pressure Groups in Britain, London, Dent.

8. For an example which concentrates on the impact of the economic recession of the 1970s, see HOOD, C. and WRIGHT, M. (eds) (1981) Big Government in Hard Times, Oxford, Martin Robertson.

9. See, for example, KING, A. (1975) 'Overload: problems of governing in the 1970s' in Political Studies, Vol.XXXIII, Nos. 2 and 3, pp 162-174 and BRITTAN, S. (1975) 'The economic contradictions of democracy' in British Journal of Political Science, Vol.5, No.1, pp 129-159. A more radical development of the same basic idea can be found in HABERMAS, J. (1976) Legitimation Crisis, Heinemann, London.

10. The locus classicus is SCHMITTER, P.C. (1974) 'Still the century of corporatism?' in *The Review of Politics*, Vol.36, No.1, pp 85-131. For more recent work with the same theme, see HARRISON, R.J. (1980) *Pluralism and Corporatism*, London, George Allen and Unwin and MIDDLEMAS, K. (1979) *Politics in Industrial Society*, London, Deutsch.

11. cf HAYWARD, J.E.S. and BERKI, R.N. (eds) (1979) *State and Society in Contemporary Europe*, Oxford, Martin Robertson. Berki uses the distinction to differentiate between the countries of Eastern Europe, on the one hand, and the United States and the countries of Western Europe, on the other. It seems to the author that its use is well-justified by the contrast between Britain and France.

12. DYSON, K. (1980) *The State Tradition in Western Europe*, Oxford, Martin Robertson, p 1. I owe a major debt to this book for drawing my attention to these basic distinctions.

13. quoted ibid. p 5.

14. cf HAYWARD, J.E.S. (1983) *Governing France: The One and Indivisible Republic*, London, Weidenfeld and Nicolson, p 21.

15. DYSON, K. (1980) op.cit. p 210.

16. See JORDAN, G. and RICHARDSON, J. (1982) 'The British policy style or the logic of negotiation' in RICHARDSON, J. (ed) *Policy Styles in Western Europe*, London, George Allen and Unwin, pp 80-110; cf BIRNBAUM, P. (1982) *La Logique de l'Etat*, Paris, Fayard, p 14, who goes so far as to suggest that the word 'state' lacks applicability in Britain because of the self-regulating quality of civil society.

17. cf HAYWARD, J.E.S. (1974) 'National aptitudes for planning in Britain, France and Italy' in *Government and Opposition*, Vol.9, No.4, pp 398-9.

18. quoted in KATZENSTEIN, P.J. (1976) 'International relations and domestic structures: foreign economic policies of advanced industrial states' in *International Organisation*, Vol.30, No.1, p 18.

19. DYSON, K. (1980) op.cit. p 218.

20. JORDAN, G. and RICHARDSON, J. (1982) op.cit. p 86.

21. DYSON, K. (1980)op.cit. p.258.

22. BIRNBAUM, P. (1982) op.cit. p 30.

23. quoted in DYSON, K. (1980) op.cit. p 271.

24. GILPIN, R. (1975) 'Three models of the future' in *International Organisation*, Vol.29, No.1, p 45.

25. For example, KINDLEBERGER, C.P. (1973) The World in Depression, 1929-1939, London, Allen Lane.

26. For example, GILPIN, R. (1976) US Power and the Multinational Corporation, London, Macmillan, pp 234-5.

27. cf KATZENSTEIN, P.J. (1976) op.cit. p 21 who refers to GERSCHENKRON, A. (1962) Economic Backwardness in Historical Perspective: A Book of Essays, Cambridge, Mass; Harvard University Press.

28. DYSON, K. (1980) op.cit. p 272; cf SULEIMAN, E.N. (1974) Politics, Power and Bureaucracy in France, Princeton, Princeton University Press, p 21.

29. HAYWARD, J.E.S. (1983) op.cit. p 172.

30. For example, GREEN, D. (1978) 'Individualism versus collectivism: economic choices in France' in West European Politics, Vol.1, No.3, pp 81-96.

31. ZYSMAN, J. (1977) 'The French state in the international economy' in International Organisation, Vol.31, No.4, p 839; cf KATZENSTEIN, P.J. (1976) op.cit. passim.

32. HAYWARD, J.E.S. and BERKI, R.M. (eds) (1979) op.cit. pp 2-3.

33. HOUSE OF COMMONS (1978) The Fishing Industry, Fifth Report of the Expenditure Committee, (Trade and Industry Subcommittee), Session 1977-1978, Vol.III, Pt.1, p 644.

34. For example, SFIA, (1983) The Administration of the Fishing Industry in France at Central and Local Level, Occasional Paper Series No.1, Fishery Economics Research Unit, Edinburgh prepared by I. Milligan, and OLIVER, T.A. (1982) The Management and Organisation of the Fish Catching Industry on the North East Coast of England, a Research Report from the Centre for Fisheries Studies at Hull College of Higher Education, esp. pp 154ff. A similarly rosy view of the French industry's organisation was expressed to the author by the former Fisheries Officer of the Transport and General Workers' Union (TGWU) - (Interview, Hull, 26.3.80.).

35. HOUSE OF COMMONS (1978) op.cit. Vol.III, Pt.1, p 667.

36. OLSON, M. (1965) The Logic of Collective Action, Cambridge, Harvard University Press, p 44.

37. cf HARRISON, R.J. (1980) op.cit. pp 13-14.

38. FINER, S.E. (1974) 'Groups and political participation' in KIMBER, R. and RICHARDSON, J.J. (eds) op.cit. pp 256-7.

39. Hence this work does not follow WILSON, F.L. (1983) 'Les groupes d'intérêt sous la cinquième république' in Revue Française de Science Politique, Vol.33, No.2, pp 220-253 and in the paper entitled 'The structures of French interest group politics' that he delivered at the 1983 Annual Meeting of the American Political Science Association. In both of these works he seeks to argue for one correct model of interest group behaviour in France

CHAPTER TWO

1. OECD annual report on fisheries quoted in Le Marin, October 15 1976.

2. Le Monde, June 26 1975.

3. Fishing News, January 10 1975.

4. Le Marin, January 22 1982.

5. Fishing News, January 28 1983.

6. The French figure was indicated to the author in an interview in Paris with the Technical Adviser to the Secretary of State for the Sea (HENNEQUIN - 28.10.83.); the British figure was given in a 1976 report of the Prices Commission entitled 'Prices and Margins in the Distribution of Fish' discussed in Fishing News, April 16 1976.

7. Le Monde, November 22 1973.
"La pêche est le seul activité de l'économie dans lequel il n'y' ait aucune relation entre le prix de revient du poisson et son prix de vente, car les mareyeurs exercent leur activité sur un marché de l'alimentation très controlé et où les aliments de substitution ou les produits d'importation ne permettent pas une répercussion mécanique des coûts".

8. BOYER, A. (1967) Les Pêches Maritimes, Paris, Presses Universitaires de France (Que Sais-Je ? No.199) p 17.

9. See article by T.Oliver in Fishing News, January 20 1984 entitled 'The ships that destroyed themselves', pp 12-14.

10. For a fuller discussion of the development of international law on fisheries and of the legal aspects of the EEC fisheries policy, see HOUSE OF COMMONS (1978) op.cit. Vol.1, Appendix 1, pp 92-118.

11. HMSO (1977) Sea Fisheries Statistical Tables 1976 and Secrétariat Général de la Marine Marchande (SGMM) (1977). Statistique des Pêches Maritimes; cf BOYER, A. (1967) op.cit.pp 33-34.

12. cf HOOD, C. (1978) 'The politics of the biosphere: the dynamics of fishing policy' in ROSE, R. (ed) The Dynamics of Public Policy, London, Sage, pp 59-79.

13. LE BIHAN, D. (1977) Organisations de Producteurs des Pêches Maritimes en France et Droit Communautaire, CNEXO, Rapports Économiques et Juridiques, No.5, p 107.

14. CARGILL, G. (1976) Blockade '75: the Story of the Fishermen's Blockade of the Ports, Glasgow, Molendinar Press, p 2.

15. Les Pêches Maritimes Françaises en 1976, a report prepared by the Direction des Pêches Maritimes of the SGMM for the Secrètariat d'Etat Auprès du Ministère de l'Equipement et de l'Aménagement du Territoire, p 27. See also internal document of SODIPEB (1981) Etat de la Flotte des Chalutiers Hauturiers du Sud-Bretagne au ler Janvier 1981 et Evolution des Flotilles au Cours de la Décennie 1970-1980, Annexe I.

16. This view was confirmed by an interviewee who worked for one of the Hull fishing companies at the time. (LIMAN - 27.3.80.).

17. HMSO (1983) Sea Fisheries Statistical Tables 1982 and CCPM (1983) Rapport sur la Production de l'Industrie des Pêches Maritimes en 1982.

18. HOOD, C. (1978) op.cit. p 76.

19. Sunday Times, 30 March 1975 and CCPM (1974) Rapports d'Activité de l'Organisation Professionnelle pour 1973, p 40.

20. Le Monde, February 19 1975.

21. CARGILL, G. (1976) op.cit. p 3.

22. ibid. pp 7-8.

23. La Vie Française, August 25 1980. The increase does not take the increase in inflation into account.

24. Information presented at the exhibition entitled 'Pêches Maritimes: Traditions et Innovations', held at the Centre Georges Pompidou, Paris, 16.6. - 20.9.82.

25. Official Journal of the European Communities (OJEC) C.205, 13.8.81. p 8.

26. CCPM (1974) op.cit. p 6; cf CHOURAQUI, G. (1979) La Mer Confisquée, Paris, Seuil, pp 133-136.

27. For a discussion of the development of EEZs at a global level, see LA DOCUMENTATION FRANÇAISE (1977) pp 66-85.

28. EEC estimates quoted in *Fishing into the 80s*, (1978) a discussion document prepared by a group chaired by J.Prescott, MP for Hull East, p 1.

29. For a fuller discussion of this issue see TROEL, J-L. (1978), *Contribution à l'Analyse de la Stratégie des Groupes Internationaux dans le Secteur des Produits de la Mer et de l'Aquaculture*, INRA, under the supervision of D. L'Hostis,(2 volumes).

30. ibid. Volume 2, pp 96-114.

31. UAP (1975) 'La crise' in *Germes* No.19, pp 2-4; cf CARGILL, G. (1976) op.cit. pp 3-4.

32. HMSO (1983) op.cit. and LA DOCUMENTATION FRANÇAISE (1980), *Rapport du Groupe de Travail: Mer et Littoral*, Préparation du Huitième Plan 1981-1985, pp 132-3.

33. *Fishing News*, August 22 1975.

34. op.cit. March 5 1976.

35. *Economist*, August 16 1980.

36. LEIGH, M. (1983) op.cit. pp 12-13.

37. ibid. p 17. National statistics suggest over 10,000 smaller vessels in France and over 6,000 in Britain.

38. WFA (1977) *Fisheries of the European Community*, Fishery Economics Research Unit, Edinburgh, Statistical Survey, Table 2.

39. ICES (1977) and (1979) *Bulletin Statistique des Pêches Maritimes*, 1974 and 1977.

40. SGMM (1977) op.cit. and 'Les pêches maritimes et les cultures marines francaises en 1982', a map with tables issued by the *Secrétariat d'état auprès du Ministère des transports chargé de la Mer*.

41. HMSO (1977) op.cit. and SGMM (1977) op.cit.

42. CONSEIL ECONOMIQUE ET SOCIAL, (1976) *L'Avenir des Pêches Maritimes*, Rapport Martray, pp 34-6.
"Les zones situés a l'intérieur de la Communauté resteront, en principe, accessibles aux flotilles de tous les Etats membres: ce qui est capital pour la France qui réalise 72% de ses captures dans les seules eaux communautaires".

43. HOUSE OF COMMONS (1978) op.cit. Vol.II, between pp 102-103, T24-6 - charts 214-6.

44. Table adapted from a BFF Ltd Press Release of December 1977 quoted in *Fishing into the 80s*, op.cit. p 3.

45. CCPM (1975) Rapports d'Activité pour 1974, p 9.

46. BLOCH, J.P. and FOURNET, P. (1977) 'Le nouveau droit de la mer et ses incidences sur les pêches maritimes françaises' in Les Cahiers d'Outre-Mer, 30(120), p 336.

47. For an example of the genre, see CCPM (1978) Rapports d'Activité pour 1977, p 181.

48. LEIGH, M.(1983) pp 168 ff.

49. cf OLIVER, T.A. (1982) op.cit. Chapter 3, 'The Agency System', pp 54-64. The issue will be discussed again in Chapter 8.

50. See, for example, HOUSE OF COMMONS (1978) op.cit. Vol.II, p 2.

51. Fishing News, November 28 1975, reported such a purchase by the Tait family of Fraserburgh, a family famous (some would say, infamous) in the industry for the size and scope of its operations.

52. Les Pêches Maritimes Françaises en 1976, op.cit. pp 4-6.

53. HOUSE OF COMMONS (1978) op.cit. Vol.IV, T21-'Representative Bodies in the Fishing Industry', p 11.

54. ibid. Vol.IV, T135, 'French and German Governmental Approaches to the Common Fisheries Policy', p 78, French figures on employment in the industry vary very widely with no one source covering the whole of the period in this study. However, there seems no reason to doubt the comparative value of the figures quoted here.

55. LA DOCUMENTATION FRANÇAISE (1970) 'Les activités maritimes du Royaume Uni, de Grande Bretagne et d'Irlande du Nord' in Notes et Etudes Documentaires No.3727, pp 33-47.

56. Interview with C.GROUHEL at Lorient (20.7.83.).

57. HMSO (1983) op.cit.

58. Official figures quoted in Le Nouvel Economiste, 26.10.81. p 76.

59. COMMISSION DES COMMUNAUTES EUROPEENNES (1979) Impact Régional de la Politique de la Pèche de la CEE - Bretagne, Informations internes sur la pêche, p 20.

60. SODIPEB (1981) op.cit. Annexe 1.

61. WFA (1978) A Case for Aid for Restructuring the Catching Sector, 3rd draft, p 28, (Confidential internal document).

62. The figures in the paragraph have been calculated on the basis of information derived from the CCPM's reports on production in the fishing industry in 1971 and 1981. The figures do not take account of the inflation rate in the intervening period.

63. The point was made in Hull by W. ROBB (26.3.80.), A. LOUGH and J.DAVIES (both 27.3.80.).

64. DEBAUVAIS, R. (1983) 'La place de l'artisanat dans la société française: l'exemple de la pêche maritime' in Dossier pour notre temps, Nos.20-21, article 1, p 3.
"Le 'risque' de l'activité est effectivement partagé entre le capital et le travail puisque la remunération de l'un et de l'autre sont uniquement fonction du produit débarqué et de sa vente."

65. Nouvel Observateur, June 13 1977.

66. The material in this paragraph comes partly from an interview with M.DION at Concarneau (20.7.83.) and partly from the same person's article entitled 'La grande pêche thonière française' which appeared in the monthly magazine La Pêche Maritime in December 1981.

67. Calculated using HOUSE OF COMMONS (1978) op.cit. Vol.II, pp 9-10. Table A.

68. SGMM (1977) op.cit. Saithe, the most caught fish, made up 85,821t. and shellfish 180,287t. of a total of 806,219t.

69. HMSO (1983) op.cit.

70. ibid. and HOUSE OF COMMONS (1978) op.cit. Vol.II, pp 9-10, Table A.

71. HMSO (1983) ibid.

72. ibid.

73. WFA (1980) The Fishing Industry: Some Facts and Figures, internal document, p.2.

74. 'Les pêches maritimes et les cultures marines françaises en 1982', op.cit.

75. ibid. cf CCPM (1983) Rapport sur la Production en 1982, p 8.

76. OREAM (1982) La Filière des Produits de la Mer dans l'Ouest, DATAR sponsored study, passim esp pp 1 and 7.

77. based on HMSO Sea Fishery Statistical Tables for the relevant years.

78. See, for example, the article 'How long to the turning of the tide?' in the September 1982 edition of the Europe magazine of the London office of the EEC Commission, pp 24-27.

79. HOUSE OF COMMONS (1978) Vol.III, Part 1, p 617.

80. Estimates given by D.AITCHISON of the SFF in Edinburgh in the course of an interview with the author (14.4.80.).

81. Le Marin, January 22, 1982.

82. Le Marin, ibid. and SGMM (1977) op.cit.

83. Le Marin, February 19 1982.

84. SGMM (1977) op.cit.

85. CCPM (1983) Rapport sur la Production en 1982, p 41.

86. Sunday Times, February 13 1983.

87. CARGILL, G. (1976) op.cit. p 8 and LA DOCUMENTATION FRANÇAISE (1978) 'La crise des industries de la pêche' in Problèmes Economiques, No.1585, pp 16-19.

88. Interview with C.GROUHEL, Lorient (20.7.83.); cf OREAM, (1982) op.cit. p 38.

89. LA DOCUMENTATION FRANÇAISE (1978) op.cit. p 16.

90. Fishing News, December 3 1976.

91. OLIVER, T.A. (1982) op.cit. Chapter 4, pp 65-72, 'The Share System'

CHAPTER THREE

1. DRISCOLL, D.J. and McKELLAR, N.(1979) 'The changing regime of North Sea fisheries' in MASON, C.M. (ed) The Effective Management of Resources, London, Frances Pinter, p 129.

2. ibid. p 135.

3. Article 6(1) of the Convention quoted ibid. p 129.

4. Article 8 of the Convention quoted ibid. p 130.

5. ibid. pp 135-8.

6. HOUSE OF COMMONS (1978) op.cit. Vol.I, p 12.

7. Fishing News, November 14 1975.

8. Interview with I. McSWEEN (14.4.80.).

9. The discussion here and subsequently is strongly influenced by RUGGIE, J.G. (1972) 'Collective goods and future international collaboration' in The American Political Science Review, Vol.66, No.3, pp 874-901.

10. ICES survey quoted in BEHRENDT, M. Revision of the Common Fisheries Policy, a paper delivered at the Conference on Technology and Challenge of the World's New Fisheries Regime, 10 June 1980. See Fishing News, June 27 1980 for full text.

11. See, for example, WATT, D.C. (1977) 'The EEC and fishing: new venture into unknown seas' in Political Quarterly, Vol.48, No.3, pp 328-336.

12. cf SHACKLETON, M. (1983) 'Fishing for a policy? The common fisheries policy of the Community' in WALLACE, H. WALLACE, W. and WEBB,C.(eds) Policy-Making in the European Community, London, John Wiley, pp 362-3.

13. LEIGH, M. (1983) op.cit. pp 72-3 and 86-7. I am endebted to this book for much of the detail in this section.

14. ibid. pp 75-6, and SHACKLETON, M. (1983) op.cit. p 363.

15. cf LEIGH, M. (1983) op.cit. pp 100-106.

16. ibid. p 104.

17. This proved to be a continuing bone of contention. cf Fishing News, December 5 and 12 1980.

18. Fishing News, May 2 1980.

19. OJEC C205, 13.8.81. p 12.

20. LEIGH, M. (1983) op.cit. pp 130-132.

21. ibid. pp 195-6.

22. ibid. p 93.

23. Le Figaro, July 1 1970.

24. OJEC 1978 No.226 Annexe, Debates of the European Parliament, February 15, p 152.

25. Regulation 2141/70 in OJEC L236, 27.10.70.

26. cf CHURCHILL, R. (1977) 'The EEC fisheries policy - towards a revision' in Marine Policy, Vol.1, No.1, p 27.

27. EEC Negotiations - Position on Fishery Limits, MAFF press release, September 14 1971.

28. See LAMBERT, J. (1974) 'The politics of fisheries in the Community' in The European Economic Community, A Post-Experience Course, The Open University, Block 3, pp 135-152.

29. See YOUNG, S.Z. (1973) Terms of Entry, London, Heinemann, p 100.

30. ibid. p 101.

31. eg LE BIHAN, D. (1977) op.cit. p 7.

32. LEIGH, M. (1983) op.cit. p 27; cf The Times, September 30 1970.

33. Le Monde, August 19 1969 and Daily Telegraph, September 30 1969.

34. LEIGH, M. (1983) op.cit. pp 32-4; cf LE BIHAN (1977) op.cit. pp 11-12.

35. LEIGH, M. (1983) op.cit. p 33.

36. EEC Council Press Notice 1892/70 (AG325) p 7; cf LEIGH, M.(1983) op.cit. p 32.

37. CCPM (1972) Rapports d'Activité en 1971, p 20. "L'entrée de la Grande Bretagne dans le Marché Commun modifiera à notre profit le regime de ses eaux, surtout après la période derogatoire de 10 ans, toutes les eaux devant alors devenir communes."

38. LEIGH, M. (1983) op.cit. pp 57-8.

39. HOUSE OF COMMONS (1978) op.cit. Vol.1, Appendix 1, p 92 - evidence of MRS BIRNIE.

40. ibid. p 106.

41. Le Monde, October 12 1972.

42. CCPM (1974) Rapports d'Activité en 1973, p 6.

43. Le Monde, November 11 1976.

44. CCPM (1975) Rapports d'Activite en 1974 pp 25-27; cf LE BIHAN (1977) op.cit. p126.

45. Le Monde, March 6 1975.

46. eg Fishing News, April 23 1976 - HAY, Chairman of the Scottish Inshore White Fish Producers' Organisation.

47. Speech to the Annual Dinner of the Federation of Fish Fryers reported in Fishing News, May 14 1976.

48. LEIGH, M. (1983) op.cit. p 82.

49. ibid. p 84; cf Fishing News, May 4 1979.

50. ibid. p 91.

51. HOUSE OF COMMONS (1978) op.cit. Vol.II, p 24 - GILLETT.

52. ibid. Vol.I, p 1206, - LAING.

53. Fishing into the 80s (1978) op.cit. p 1.

54. WFA (1977) op.cit. p 7 - J. REGNIER, Editor, France Pêche

55. LEIGH, M. (1983) op.cit. p 169.

56. Le Monde, September 17 1979.

57. cf CCPM (1978) Rapports d'Activité en 1977, p 8.

58. ibid. p 15, and same document for 1978, p 6.

59. La Pêche Maritime, September 20 1980, p 494. "...une refonte de l'organisation commune des marchés dont les mecanismes, arrêtés en 1970 dans un contexte de relative abondance, se révèlent de moins en moins bien adaptés à la situation de pénurie que connaît la pêche communautaire...."

60. CCPM (1982) Rapports d'Activité en 1981, p 4. The measures were not finally approved until January 1983; cf LEIGH, M. (1983) op.cit. pp 114-5.

61. LEIGH, M. (1983) op.cit. p 109.

62. HANSARD (1982-3) Vol.35, Col.909 - WALKER.

63. ibid. Col.905.

64. LEIGH, M. (1983) op.cit. p 87.

65. ibid. p 106

CHAPTER FOUR

1. MORDREL, L. (1972) Les Institutions de la Pêche Maritime, Thèse pour le Doctorat en Droit, Paris II, provides an excellent analysis of the impact of the past on present institutions.

2. Interview with J.C. HENNEQUIN, Paris (28.10.83.)

3. MORDREL, L. (1972) op.cit. Vol.I, pp 54-7.

4. HOUSE OF COMMONS (1978) op.cit. Vol.III, Pt.1, p 665- Skipper NIELSON.

5. Interview with R. KEMP, Hull, (26.3.80).

6. HOUSE OF COMMONS (1978) op.cit. Vol.II, p 313 - HAMLEY.

7. For a statement in English of the arrangements, see HOUSE OF COMMONS (1978) op.cit. Vol.IV, p 82

8. CONSEIL ECONOMIQUE ET SOCIAL (1976) L'Avenir des Pêches Maritimes, presented by J.Martray, p 189.

9. Interview with A.LOUGH, Hull, (26.3.80.).

10. The information on 'administrateurs maritimes' comes chiefly from my interview with J.C. HENNEQUIN, (28.10.83.).

11. MORDREL, L. (1972) op.cit. p 157 ff and LE BIHAN, D. (1977) op.cit. pp 26-28

12. For the complete text of the 1945 Decree and its subsequent amendments, see CCPM (1982) Statuts de l'Organisation Professionnelle des Pêches Maritimes, May 1982; for a briefer account of the French structure in English, see SFIA (1983) op.cit. which is in itself an update of an earlier paper WFA (1972) Industry Participation in the Administration of the Fishing Industry in France, No.72/3, Fishery Economics Research Unit, Edinburgh, prepared by A.D. Insull and D.A. Palfreman.

13. CCPM (1982) op.cit. p 19.

14. ibid p 24.

15. ibid. pp 24, 30 and 31.

16. ibid. pp 9 and 10.

17. For a brief account of the nature of these two bodies see HOUSE OF COMMONS (1978) Vol.II, p 4: for a fuller understanding see their annual reports up to the time they were merged in 1981; and for a French view see LA DOCUMENTATION FRANÇAISE (1970) 'L'industrie des pêches brittaniques' in 'Les activites maritimes du Royaume Uni, de Grande Bretagne et d'Irlande du Nord', Notes et Etudes Documentaires, No.3727, pp 37-47.

18. Interview with W. ROBB (26.3.80.).

19. The author was given a glimpse of the room containing all these dossiers!

20. For full details see CCPM (1982) op.cit. pp 6-8 and 65-113.

21. ibid. p 7.

22. ibid. pp 4-6.

23. For a wry comment on the worth of these committees see HOUSE OF COMMONS (1978) op.cit. Vol.III, Pt.2, p 1306.

24. HAYWARD, J.E.S. (1983) op.cit. p 70.

25. MORDREL, L. (1972) op.cit. p 360. See also BIDET, J. (1974) 'Sur les rapports d'être de l'idéologie, les rapports sociaux dans le secteur de la pêche' in La Pensée, 174, pp 53-66.

26. Interview with D. AITCHISON (14.4.80.).

27. HOUSE OF COMMONS (1978) op.cit. Vol.II, p 315.

28. ibid. Vol.II, p 26 - GILLETT, Fisheries Secretary, DAFS, 1971-6.

29. Interview with D. CAIRNS, Hull, (26.3.80.).

30. Private interview.

31. HOUSE OF COMMONS (1978) op.cit. Vol.II, p 25.

32. ibid. Vol.I, p 53, para 136.

33 NFFO Newsletter, November 16 1979.

34. Fishing News, May 21 1982.

35. HANSARD (1976) Vol.905, Col.1572. An unfavourable view of MAFF was repeated in interviews with D.CAIRNS (26.3.80.) and R.KEMP (26.3.80).

36. Interview with I. McSWEEN (14.4.80) who was also much more enthusiastic about DAFS.

37. ibid. referring to career of one of DAFS officials.

38. HANSARD (1975), Vol.889, Col.1243.

39. HOUSE OF COMMONS (1978) Vol.II, p 337.

40. Private interview.

41. Interview with W. ROBB, (26.3.80.).

42. Information provided by S.VICKERS at present completing his thesis on the British fishing industry during the 1970s and early 1980s.

43. HOUSE OF COMMONS (1978) op.cit. Vol.II, p 119 ff.

44. Fishing News, January 31 1975.

45. Lyon Dean of the HIB made this point in HOUSE OF COMMONS (1978) op.cit. Vol.II, p 230.

46. ibid. Vol.II, p 131.

47. CONSEIL ECONOMIQUE ET SOCIAL (1976) op.cit. p 4.

48. ibid. p 185,
'...tribune aux revendications (souvent contradictoires) des diverses catégories représentées'.

49. ibid. pp 175 ff.

50. Le Monde, February 21 1975 and January 19 1976.

51. cf Note d'information sur le FIOM, May 1981, p 1.

52. Interview with M.LE BELLER (23.7.82).

53. La Vie Française, April 24, 1975.

54. Le Marin, June 1 1979
'L'action du FIOM ... constitue un facteur d'entraînement, d'accompagnement, d'une volonté, au plan local, de constituer un système d'assurances face aux aléas des prises'

55. CCPM (1975) Rapports d'Activite en 1974, p 35.

56. CCPM (1977) Rapports d'Activite pour 1976 - Rapport du FIOM, p 2. After this year, FIOM's report ceased to be bound with that of the CCPM.

57. Interview with Y.L'HELGOUACH, Concarneau (19.7.83.).

58. Interview with A.GRUENAIS, Paris, (15.4.83.).

59. Interview wih J.C. HENNEQUIN (28.10.83.).

60. eg Le Monde, January 19-20 1975.

61. 'L'organisation maritime' in Les dossiers de l'histoire No.20, July-August 1979,'La Mer', p 108 and Le Marin, August 4 1978.

62. CCPM (1980) Rapports d'Activité pour 1979, p 28; cf Le Monde, August 2 1978.

63. CCPM (1981) Rapports d'Activité pour 1980 p 2.
"...chacun y voit l'espoir longtemps nourri d'être enfin entendu, compris, perçu par un Pouvoir accessible, ouvert et attentif - Elle est là promesse,cette novation, de tous les redressements attendus depuis 1975, la promesse d'une pêche à nouveau rentable, promesse en tous les cas que cette activité économique essentielle et son contexte humain ne seront pas sacrifiés à des considérations statistico et technocrato-économiques désincarnées".

64. Le Marin, May 28 1982.

65. CCPM (1983) Rapports d'Activité pour 1982 p 2.

CHAPTER FIVE

1. HAYWARD, J.E.S. (1976) 'Institutional inertia and political impetus in France and Britain' in European Journal of Political Research, Vol.4, No.4, p 347.

2. OCDE (1981) Examen des Pêcheries dans les Pays Membres de l'OCDE, Paris, p 3.
"Dans bon nombre de pays de l'OCDE, il est habituel que l'industrie de la pêche reçoive des aides financières de l'Etat mais il faut noter que les aides se sont intensifiées ces dernières années".

3. HOOD, C. (1978) op.cit. pp 59-63.

4. LA DOCUMENTATION FRANÇAISE (1970) op.cit. p 47.

5. Le Monde, November 22 1973.

6. Fishing News, July 18 1975.

7. Le Monde, February 21 1975.

8. HOUSE OF COMMONS (1978) op.cit. Vol.II, p 19; OREAM (1982) op.cit. Annexe 2 and Rapport Annuel du FIOM, 1982, Annexe No.3. These figures do not take account of inflation.

9. OREAM (1982) ibid; Fishing News, March 21 and August 8 1980.

10. Fishing News, April 3 1981 and October 29 1982.

11. OREAM (1982) op.cit. 'L'Enjeu', p 3 and Annexe 2.

12. I am grateful to Remy Debeauvais of CEASM for drawing my attention to this point.

13. eg OJEC C205/8, 13.8.81.

14. Fishing News, August 22 1980.

15. Fishing News, July 18 1980.

16. LA DOCUMENTATION FRANÇAISE (1980) op.cit. p 76.
"La France doit disposer d'un outil spécifique permettant de satisfaire l'aspect qualitatif de la demande (produits frais), aussi bien au plan de la production que de la distribution."

17. ibid. p 77.
"...le maintien d'une activité halieutique suffisante doit avoir un rôle dans la limitation du deficit de la balance."

18. CLUB SOCIALISTE DU LIVRE, (1981) La Mer Retrouvée, Paris, pp 13 and 14.
"une composante indispensable de la politique de l'emploi et de l'amenagement du territore des régions cotières ... un chaînon important de nos ressources alimentaires."

19. WFA (1978) A Case for Aid for Restructuring the Catching Sector, Edinburgh, Fishery Economics Research Unit, p 2; cf Fishing News, August 15 1980 where Meek reiterates this case.

20. HMSO (1975) Food from our own resources, Cmnd 6020 referred to in an interview with T.LIMAN (27.3.80).

21. HANSARD (1975) Vol.889, Col.1231.

22. Fishing News, February 22 1980.

23. CCPM (1978) Rapports d'Activité pour 1977, p 15.

24. WFA (1978) ibid. p 3.

25. HOOD, C. and WRIGHT, M. (eds.) (1981) op.cit. pp 4-5.

26. Le Nouvel Journal, June 30 1978.

27. Fishing News, August 8 1980.

28. OREAM (1982) op.cit. Annexe 2.

29. Fishing News, September 12 1980.

30. CCPM (1974) Rapports d'Activité pour 1973.

31. Les Echos, February 21 1975.

32. Le Figaro, April 23 1976.

33. L' Express, August 16 1980.

34. CCPM (1982) Rapports d'Activité pour 1981, p 24.

35. Fishing News, July 4 and August 1 1975.

36. Fishing News, October 31 1975.

37. eg Fishing News, February 6 1976.

38. eg Fishing News, August 15 1980.

39. Fishing News, March 5 1982; cf La Pêche Maritime, March 1982, p 128.

40. CCPM (1972) Rapports d'Activité pour 1971 p 41.
"Le premier reflexe de l'armement fut de se tourner vers les pouvoirs publics."

41. UAP (1975) op.cit. p 12.
"Seule, la puissance publique a vocation à relancer l'activité halieutique comme elle a fait dans d'autres branches - pour protéger l'emploi et les capacités de production."

42. CCPM (1978) Rapports d'Activité pour 1970, p 25.

43. CCPM (1981) Rapports d'Activité pour 1980, p 27.

44. Interview with J. PLORMEL, Boulogne (30.9.83.) who thought the new government to be particularly favourable to the industry, despite his own political disagreement with it.

45. Fishing News, May 9 1976.

46. Fishing News, March 21, 1980.

47. eg Fishing News, August 8 1980.

48. HANSARD (1974) Vol.880, Or.1232-3.

49. Fishing News, August 8 1980.

50. For an interesting discussion of this issue, see interview with Silkin in Fishing News, May 25 1979.

51. Fishing News, August 8 1975.

52. Fishing News, August 22 1980.

53. CCPM (1975) Rapports d'Activité pour 1974, p 30.

54. FOURNET, P. (1977) 'La grande pêche morutière française' in La Pêche Maritime, May 1977, pp 271-277; cf Témoignage Chrétien, September 4 1975 and Le Marin, December 29 1978 for earlier and later views of the problems of this part of the industry.

55. BLOCH, J.P. and FOURNET, P. (1977) op.cit. pp 341-3.

56. eg La Pêche Maritime, September 1978 p 523.

57. Decree 78-144 of February 3 1978, published in the Official Journal of February 11.

58. eg Le Monde, October 12 1972, discussing Pompidou's meeting with the leaders of the fishing industry well before UNCLOS III started.

59. La Pêche Maritime, July 1977, p 391.
"La France a sous sa juridiction, onze millions de kilomètres carrés de mer, ce qui la place au troisième rang dans le monde ... La mer est une nouvelle frontière de la France."

60. La Pêche Maritime, April 1982, p 180.

61. FOURNET, P. (1978) 'La crise des pêches maritimes britanniques' in La Pêche Maritime, October 1978, p 574. "...les 200 milles offrent au Royaume - Uni la possibilité de contrôler de vastes espaces maritimes et de redéployer sa flotte de pêche lointaine sur plusieurs océans."

62. Fishing News, April 16 1982.

63. DUHAMEL, G. and HUREAU, J.C. (1981) 'La situation de la pêche aux îles Kerguelen en 1981' in La Pêche Maritime, May 1981 pp 272-9.

64. HMSO (1982) Falkland Islands Economic Study 1982, Cmnd 8653, Chairman - The Rt.Hon. Lord Shackleton, p 12.

65. JOHNSON, B. and ZACHER, M.W. (eds) Canadian Foreign Policy and the Law of the Sea, Vancouver, University of British Columbia Press, pp 137-9.

66 HOUSE OF COMMONS (1978) op.cit. Vol.I, p 26, para 68.

67. Fishing News, April 28, 1978.

68. HMSO (1982) op.cit. p 18.

69. Fishing News, March 2 1984.

70. LEIGH, M. (1983) op.cit. p 109.

71. CCPM (1975) Rapports d'Activité pour 1974, p 7.

72. For the details in this paragraph, see Le Monde, August 17 1979 and Le Télégramme (de Brest), October 17/18 1981.

73. Le Monde, December 25 1979.

74. Voies, June/July 1979, p 10. (This was the journal of the Transport Ministry). "Quoi qu'il en soit l'expérience du Jutland revêt un caractère un peu symbolique. Elle est en effet le témoignage de la volonté de la pêche francaise de faire face aux contraintes d'accès à la ressource. Elle marque aussi la volonté des Pouvoirs Publics d'accompagner les initiatives des professionels."

75. Interview with LE BELLER (23.7.82) and CCPM (1981) Rapports d'Activité pour 1980, p 37.

76. Quoted from the company's publicity in Le Télégramme, October 17/18 1981. "pour assurer l'indépendance de la Pêche Maritime Française."

77. Le Monde, September 23 1981, - "le trésor des Kerguelens"

78. La Pêche Maritime, May 1982, p 249.
"il faut que la pompe reste amorcée pour le jour où toute la flotte se déploiera aux Kerguelens"

79. Le Monde, November 11 1982.

80. Le Marin, May 20 1983.

81. La Pêche Maritime, October 1981, pp 555-6.

82. Le Marin, February 26 1982.

83. Interview with M.DION, Concarneau (20.7.83.).

84. Le Figaro, September 18 1979.

85. La Pêche Maritime, February 1982.

86. La Pêche Maritime, December 1981.

87. Interview with M.DION (20.7.83.).

88. Le Marin, October 1 1982.

89. Interview with M.DION (20.7.83.) and LEIGH, M. (1983), op.cit. p 142.

90. Fishing News, March 21 and May 2 1975.

91. Fishing News, October 15 1976.

92. I am grateful to Ian Scott of the SFIA for drawing my attention to this comparison.

93. Fishing News, January 11 and March 7 1980.

94. Fishing News, March 21 and 28 and August 22 1980. In his interview with me, E.OAKESHOTT, (15.4.80.) argued that no fish were unexploited that it was economic to exploit: this assumes that the state does not seek to intervene to influence profitability as FIOM did.

95. HOUSE OF COMMONS (1978) op.cit. Vol.III, Pt 2, p 1309.

96. La Pêche Maritime, March 1977, p 167; April 1977, p 226; May 1977, p 287.

97. This information was provided in private correspondence by Ian Scott of the SFIA. The initiative is specifically (and unfavourably) compared with COFREPÊCHE's activities in Fishing News editorial of March 20 1981.

98. The quote, from PEYREFITTE, A. (1976) Le Mal Français, Paris Plon, is quoted in La Pêche Maritime, August 1977, p 466.

99. Parrès, Head of the UAP, in La Pêche Maritime, December 1977, p 716.

100. Le Marin, July 25 1980.

101. Le Monde, September 23 1981.

102. Ouest France, April 19 1982.

103. Interview with LE BELLER (23.7.1982.).

104. La Pêche Maritime, May 1982, p 249.

105. eg interview with T. LIMAN (27.3.80.).

106. The phrase 'ils jouent tous le jeu' was used by LE BELLER in his interview with me (23.7.1982.).

107. HAYWARD, J.E.S. (1982) 'Mobilising private interests in the service of public ambitions: the salient element in the dual French policy style?' in RICHARDSON, J. (ed) op.cit p 116.

CHAPTER SIX

1. Economist, August 16 1980, p 40.

2. WFA (1977) op.cit. Statistical Tables, Table 3.

3. EUROSTAT (1979) Basic Statistics of the Community.

4. LE MONDE (1981) 'L'Election Présidentielle 26 avril- 10 mai 1981', Supplément aux Dossiers et Documents du Monde, May 1981, p 128. For a diluted version of Crosland's 'off the record' comment, see VOLLE, A. and WALLACE, W. (1977) 'How common a fisheries policy?' in The World Today, Vol.33, No.2, p 72.

5. See, for example, HAYWARD, J.E.S. (1983) op.cit p 84 ff and WRIGHT, V. (1978) The Government and Politics of France, London, Hutchinson, Chapter 5, pp 107-123.

6. cf WILSON, F. (1979) 'The revitalization of French parties' in Comparative Political Studies, Vol.XII, No.1, pp 82-103.

7. Le Marin, March 27 1981, quoting the article of P. Herry, which appeared in the January/February copy of Echanges Internationaux in the same year.
"Rattachés au ministère des Transports entre Concorde et Airbus les professionels de la mer sont pour leur administration de "tutelle" des marins avant d'être des producteurs. Le CCPM, né sous Vichy, dote le secteur d'une organisation professionelle corporative. Les syndicats s' y engluent, coincés entre les armateurs, les mareyeurs, les poissoniers et l'administration."

8. Le Marin, ibid.

9. Interviews with Y. GUILLEMONT (19.7.83.) and M.DION (20.7.83.).

10. Interviews with A. GRUENAIS (15.4.83.) and Y. L'HELGOUACH (19.7.83.) - The former's actual phrase in referring to the exclusion of controversial subjects from the CCPM was 'la vie serait imbuvable'.

11. BOARDMAN, A. (1976) 'Ocean politics in Western Europe' in JOHNSTON, D.M. (ed) Marine Policy and the Coastal Community, London, Croom Helm, p 192 and YOUNG, S.Z. (1973) op.cit. p 99.

12. GILCHRIST, A. (1978) Cod Wars and How to Lose Them, Edinburgh, Q Press, p 64.

13. Interview with P.SOISSON, Paris, (1.10.80.).

14. Interviews with D.AITCHISON (14.4.80.) and D.CAIRNS (26.3.80.).

15. SFF newsletter no.2 (undated) p 5.

16. Fishing News, March 13 1981.

17. For the fullest presentation of all these debates up to 1980 see DADRE, V.C. (1980) Vers une Politique Commune de la Pêche dans la Communauté: Fondements Juridiques et Problèmes d'Élaboration, Thèse pour le Doctorat du 3e cycle, Université de Bordeaux, Faculté de Droitet des Sciences Economiques.

18. HANSARD (1977-78) Vol.951, Col.1255.

19. HANDSARD (1982-83) Vol.35, Col.906.

20. Le Marin, June 10 1977.

21. Le Monde, August 27 1980.
"le vide de l'hémicycle n'est rompu que par une dizaine de deputés ou senateurs luttant contre le marchand de sable."

22. HANSARD (1977-78) Vol.951, Col.1215.

23. Economist, January 21 1978.

24. Fishing News, April 23 1976.

25. Interview with D.AITCHISON (14.4.80.).

26. L'Humanité, October 11 1976 and May 17 1980.

27. Le Marin, November 3 1976 and La Pêche Maritime, March 1979.

28. Le Monde, May 15 1981.

29. cf Fishing News, July 23 1976.

30. HANSARD (1980-81) Vol.995, Col.628; cf Fishing News, December 5 1980.

31. Fishing News, May 15 and 29 1981.

32. Le Marin, December 19 1980 and February 13 1981.

33. For articles recording the struggles of life at sea see, for example, Libération, December 30 1975 and Le Monde May 19 1976. Grosrichard, the Le Monde fishing correspondent, wrote many good pieces over the period, eg August 26 1980.

34. Interview with P. SOISSON, (1.10.80.).

35. Interview with B. DUBREUIL, (11.5.83.).

36. Private interview. For an amusing account of the atmosphere in Brussels, see Fishing News, February 20 1981.

37. CANETTI, E. (1981) Crowds and Power, Harmondsworth, Penguin, p 200.

38. The idea that the character of the issue at stake heavily influences the way it is handled can be seen in LOWI, T. (1972) 'Four systems of policy, politics and choice' in Public Administration Review, July - August 1972, pp 298-310. He offers a distinction between 'distributive' and 'redistributive' issues which fits the French and British definitions of the fisheries' problem. For further discussion of the distinction in the Community context, see WALLACE, W. (1983) 'Less than a federation, more than a regime: the Community as a political system' in WALLACE, H. WALLACE, W. and WEBB, C. (eds) op.cit. p 421.

39. cf IONESCU, G. (1975) Centripetal Politics, London, Hart-Davis MacGibbon.

40. BOARDMAN, R. (1976) op.cit. p 193.

41. Interview with D. AITCHISON (14.4.80.); cf HOOD, C. (1973) 'British fishing and the Iceland saga' in Political Quarterly, Vol.44, No.3, p 351.

42. eg Le Monde, October 14-15, 1973.

43. CCPM (1974) Rapports d'Activité pour 1973 p 44.

44. GILCHRIST, A. (1978) Cod Wars and How to Lose Them, Ediburgh, Q Press p 110; cf LEIGH, M. (1983) op.cit. p 66.

45. ibid. p 106.

46. See HOUSE OF COMMONS (1978) op.cit. Vol.I, p 86, para 267 for the choice of Hattersley; as for Laing his association with the failure was referred to in my interview with J.DAVIES (27.3.83.).

47. cf Guardian, May 29 1979.

48. Interview with D. CAIRNS (26.3.80.); cf HOUSE OF COMMONS (1978) Vol.III, pp 651-660.

49. ibid.

50. Interview with T. LIMAN, (27.3.80.); cf 'The Humberside Fishing Industry', Report for HM Government, 11 June 1976 p 11, para 6.2.

51. HANSARD (1974) Vol.873, Wr 497 and (1974-5) Vol.882, Wr 271.

52. HOUSE OF COMMONS (1978) op.cit. Vol.II, p 1110.

53. Fishing News, April 22 1983.

54. Fishing News, May 6 1983.

55. Le Marin, December 8 1978.

56. Union Fédérale Maritime: CFDT - 'Rapport Général: Activités 1975-78' pp 5 and 6.

57. Article 117 states: 'Member states agree upon the need to promote improved working conditions and an improved standard of living for workers, so as to make possible their harmonisation while the improvement is being maintained'.

58. Interview with A. GRUENAIS (15.4.83.).

59. Private interview.

60. Fishing News, December 17/24 1982 and throughout January and February 1983.

61. Fishing News, October 17 1975 and interview with I. McSWEEN (14.4.80.).

62. Fishing News, April 2 1976.

63. The following information was provided in my interview with R. KEMP (26.3.80.).

64. Le Monde, January 7 1971.

65. Le Marin, February 26 1982, referring back to July 1978.

66. Le Marin, October 8 1982.

67. HANSARD (1982-83) Vol.35, Cols.905-913.

68. Interview with J. DAVIES (27.3.80.) and LEIGH, M. (1983) op.cit. p 66.

69. SFF newsletter No.1 (undated) p 1.

70. Daily Telegraph, September 11 1980; cf HOUSE OF LORDS (1980) EEC Fisheries Policy, 67th Report of the Select Committee on the European Communities, Session 1979-80.

71. MACKINTOSH, J.P. The Impact of EEC Policies on British Fishing and Agriculture, European Movement (undated but before British Accession), p 7.

72 DADRE, V.C. (1980) op.cit. p 247 - she argues that the Irish did become aware of this possibility when they abandoned their demand for an exclusive 50 mile limit at the beginning of 1978.

73. HANSARD (1980-81) Vol.995, Col.639.

74. Private interview confirmed in discussion with A. MITCHELL, MP for Grimsby at the UACES conference at Keele in January 1984.

75. Fishing News, May 4 1979.

76. Conversation with A. MITCHELL, MP for Grimsby at UACES conference at Keele in January 1984.

77. ARDAGH, J. (1977) The New France, Harmondsworth, Penguin, p 212.

78. La Pêche Maritime, February 1977, p 67.
"J'ai demandé aux ministres des Transports et des Affaires étrangères de se montrer intransigeants dans la défense des droits français de la pêche. Les marins n' ont pas d'inquiétude à se faire. Leurs droits traditionnels de pêche seront reconnus et protégés."

79. L'Humanité, July 25 1979, and interview with L'HELGOUACH (19.7.83.).

80. CCPM (1980) Rapports d'Activité pour 1979, p 9.

81. Le Monde, September 19 1979.

82. Daily Telegraph, September 21 1979.

83. CCPM (1980) op.cit. p 11.
"protéger ses ressortissants et leur réimbourser les préjudices subis lorsque les intéressés se trouvent être en règle avec les dispositions arrêtées en la matière par la France et approuvées par la Commission de Bruxelles."

84. ibid. p 14.

85. Guardian and Financial Times, September 28 1979.

86. cf HAYWARD, J.E.S. and BERKI, R. (eds) (1979) op.cit. p 237.

87. quoted ibid. p 54.

88. HAYWARD, J.E.S. (1976) op.cit. p 34.

CHAPTER SEVEN

1. Fishing News, August 13 1976.

2. See La Croix, August 20 1970 (Lorient), Le Monde, October 16 and 18 1971, (Île d'Oléron) and Le Monde, December 5 1973, (Martiques and Fos).

3. RICHARDSON, J. (ed.) (1982) op.cit. p 204.

4. Le Monde, January 19-20 1975.

5. Le Monde, December 23 1974.

6. Le Monde, February 19 1975. The material for the account that follows is derived from the press cuttings library in the Fondation Nationale des Sciences Politiques in Paris.

7. HANSARD (1975) Vol.889, Col.1275.

8. HANSARD ibid. Cols.1237-8.

9. CARGILL, G. (1976) op.cit. p 16. The details of what occurred in 1975 are derived from this book, with some additional material from the press cuttings library at Chatham House.

10. eg The Times, April 2 1976.

11. Fishing News, August 1 1980, and February 6, 13 and 20 1981.

12. The press coverage of the period points to four further incidents in 1975, one in 1976, four in 1977, three in 1978 and one in 1979. This same press coverage provides the bulk of the material in the account that follows. For a good discussion of the dispute from a rather different point of view, see EISENHAMMER, J. (1983). The Parti Communiste Française And the Confédération Générale du Travail in Contemporary French Politics: A Study of Some Aspects of the Organizations and Their Relationship, D.Phil Thesis, Hilary Term 1983, Nuffield College, University of Oxford, Chapter Six.

13. <u>Le Point</u>, November 3 1980 - ESSIG, Director-General of the DGMM, who was sacked at the end of the year for his involvement in the debacle.

14. The second example is taken from a French case study of the dispute: CORLAY, J.P. (1983) 'La grève des marins-pêcheurs de l'été 1980 en Basse Normandie. Géographie d'un conflit social en milieu halieutique' in <u>Façade Atlantique - Les Faits d'Occupation Conflictuelle du Littoral</u>, Université de Nantes, Groupe de Recherche sur les Structures Economiques et les Rapports Sociaux, p 207.

15. <u>Le Monde</u>, August 19 1980, indicated that interpretations were divided between those who saw in the operation the Dunkirk spirit and those for whom it brought back memories of Agincourt.

16. <u>Le Monde</u>, August 17-18 1980.

17. quoted in <u>Quotidien de Paris</u>, August 28 1980. "...confirmer sa directive de maintien d'ouverture au trafic des grands ports maritimes françaises dont l'activité n'est pas concernée par les problèmes de la pêche."

18. <u>Ouest France</u>, September 3 1980, <u>Le Figaro</u>, September 4 1980 and <u>La Croix</u>, September 13 1980.

19. <u>Le Monde</u>, September 19 1980.

20. CARGILL, G. (1976) op.cit. p 31.

21. <u>Guardian</u>, March 31 1975.

22. CARGILL, G. (1976) op.cit. p 33.

23. <u>Fishing News</u>, April 23 1976.

24. CARGILL, G. (1976) op.cit. pp 33-4.

25. CCPM (1981) <u>Rapports d'Activité pour 1980</u>, pp 25-27.

26. <u>Ouest France</u>, August 19 1980. "En tant que chef d'entreprise, il n'est jamais intéressant de faire face à une grève, mais aujourd'hui, en tant qu'armateur, nous comprenons parfaitement la réaction des marins et la decision qu'ils ont prise de manifester en accord avec leurs collègues des ports du Nord. Voilà plus de deux ans, que l'armement réclame des mesures particulières et des ratifications spéciales du carburant destiné à la pêche."

27. EISENHAMMER, J. (1983) op.cit. p 279.

28. <u>Libération</u>, August 18 1980.

29. CARGILL, G. (1976) op.cit. p 39.

30. ibid p 40.

31. Fishing News, January 18 1980.

32. NFFO Newsletter, May 1980.

33. Fishing News, February 6 1981 - LOVIE of SFO LTD.

34. ibid. February 20 and 27 1981.

35. ibid. February 8 and March 7 1980.

36. CCPM (1981) op.cit. pp 23-4.
"...si d'aucuns sont d'accord sur l'illégalité de plusieurs actions menées par les pêcheurs, ils ne reconnaissent pas moins à celles-ci une part de légitimité en raison du profond désarroi de la pêche française."

37. EISENHAMMER, J. (1983) op.cit. pp 290 and 301.

38. Fishing News, April 9 1976 - BARR, Candidate for Inverness.

39. L'Humanité, August 13 1980.

40. Le Monde, August 27 1980.
"75% de nos prises sont faites dans les eaux étrangères. La France ne peut refuser l'Europe de la pêche."

41. CORLAY, J.P. (1983) op.cit. p 232.

42. Le Monde, September 5 1980.

43. Le Figaro, August 11 1980.

44. The Times, March 25 1975.

45. Guardian, April 1 1975.

46. Le Monde, September 19 1980.

47. Le Marin, July 3 1981.
"Le temps du mépris et des guerres navales est passé".

48. eg Le Monde, August 9 1980.

49. Le Monde, September 3 1980.

50. Contrary to the argument of EISENHAMMER, J. (1983) op.cit. p 272-3.

51. cf HAYWARD, J.E.S. (1983) p 64.

52. HANSARD (1975) Vol.889, Col.29.

53. HANSARD (1975) Vol.889, Col.1232.

54. ibid. Col.1233.

55. CCPM (1981) Rapports d'Activité pour 1980, p 27. "si le gouvernement avait voulu prêter une oreille plus attentive aux multiples appels lancés par les professionels."

56. cf CORLAY, J.P. (1983) op.cit. p 232 - at Port-en-Bessin, a majority still voted for Giscard in the 1981 election.

57. CARGILL, G. (1976) op.cit. p 49.

58. Fishing News, April 11 1975.

59. Fishing News, February 6, 20 and 27 1981.

60. Fishing News, June 5 1981.

61. CCPM (1975) Rapports d'Activité pour 1974, p 30.

62. See, for example, SCHWARTZENBERG, R.G. (1981) La Droite Absolue, Paris, Flammarion, pp 246-7.

63. Interview with J.C. HENNEQUIN (20.10.83.).

64. Le Monde, August 14 1980 and La Croix, August 15 1980.

65. EISENHAMMER, J. (1983) op.cit. p 273 fn.

66. Le Monde, September 4 1980.

67. eg Quotidien de Paris, August 22 1980.

68. Le Monde, December 14-15 and 23-24 1980.

69. Original plan reported in Le Monde, August 8 1980; interpretation given by CGT official LAGAIN in Libération, August 14 1980; interpretation accepted by Economist, August 23 1980 and New York Times, August 25 1980.

70. cf HAYWARD, J.E.S. (1983) op.cit. p 67.

CHAPTER EIGHT

1. WATT, D.C. (1977) op.cit. p 335.

2. Interview with W. ROBB (26.3.80), who regularly accompanied deep-sea captains to sea.

3. Guardian, October 29 1979.

4. SGMM (1977) Les Pêches Maritimes Françaises en 1976 op.cit. p 9. "Malheureusement, beaucoup de patrons et d'armateurs n'ont pas encore compris que l'avenir de leur profession était lié à la qualité des informations qu'ils devraient fournir et il est à craindre que leurs droits ne puissent être efficacement défendus dans les années qui viennent."

5. HOUSE OF COMMONS (1978) op.cit. Vol.III, Pt II, p 1298 - LEE.

6. The author observed this in a trip to sea with a captain from Loctudy in South Finistère.

7. COMMISSION OF THE EUROPEAN COMMUNITIES (1976) Forms of Cooperation in the Fisheries Sector, Brussels, Information on Agriculture. No.9, pp 1-28 and COMMISSION DES COMMUNAUTES EUROPEENNES (1970) Formes de Coopération dans le Secteur de la Pêche, Bruxelles, Informations internes sur l'agriculture, No.68 pp 1-20 and No.69 pp 1-100 provide the bulk of the information in the rest of this section.

8. HOUSE OF COMMONS (1978) op.cit. Vol.III, Pt.II, p 832 - EDMUNDS.

9. Interview with J. PLORMEL (30.9.83).

10. COMMISSION OF THE EUROPEAN COMMUNITIES (1976) op.cit. p 39 and pp 93-95.

11. Fishing News, July 11 1975.

12. ibid. July 25 1975.

13. LAING, A. (1971) 'The Common Fisheries Policy of the Six' in Fish Industry Review, Vol.1, No.2, p 8.

14. LE BIHAN (1977) op.cit. pp 1 and 14-16.

15. CCPM (1972) Rapports d'Activité pour 1971, p 4. "La conception communautaire des organisations de producteurs est, à nos yeux, irréaliste, parce que ces organisations ne sont ni obligatoires, ni interprofessionelles."

16. LE BIHAN (1977) op.cit. p 92.

17. ibid. p 66.

18. La Vie Française, April 24 1975.

19. ibid. February 13 and November 12 1976.

20. Le Marin, June 20 1980.

21. Interview with M. CLAIROUIN (18.11.83) who provided much of the information in the following paragraphs.

22. Fishing News, March 5 1982.

23. Le Marin, May 23 1980.

24. Fishing News, April 30 1982.

25. ibid. July 24 1981.

26. eg ibid. May 28 1982.

27. ibid. April 24 1981.

28. Le Marin, September 17 1976.

29. SFIA (1983) op.cit. p 15 and interview with C. GROUHEL, (20.7.83.).

30. Interview with I. McSWEEN (14.4.80.).

31. Le Marin, June 24 1977. For futher details of SCOMA and the cooperative at St.Guenolé see Le Marin, June 23 1978.

32. Fishing News, July 25 1982.

33. OLIVER, T.A. p 10 of chapter on Fish Producer Organisations in work on British fishing cooperatives as yet unpublished.

34. Fishing News, January 25 1980 - LOVIE.

35. ibid. July 18 1980 - BOZMAN of Hull.

36. cf TROEL, J-L. (1978) op.cit. Tome 2, pp 96-114.

37. LA DOCUMENTATION FRANÇAISE (1978) op.cit. p 18.

38. Le Marin, October 9 and December 4 1981.

39. OLIVER, T.A. (1982) op.cit. Chapter 3 'The Agency System', pp 54-64 is the basic source for the folowing paragraphs. His work relates to the North East Coast of England.

40. ibid. p 56.

41. ibid. p 58.

42. LA DOCUMENTATION FRANÇAISE (1970) op.cit. p 43.

43. Le Monde, April 4 and May 8 1975.

44. Le Marin, April 14 1978.

45. Interview with J.C.HENNEQUIN (20.10.83.).

46. ibid. cf Le Marin, April 5 1980.

47. Guardian, October 26 1979.

48. Le Marin, November 16 1979.

49. ibid. November 20 1981.

50. Le Marin, January 9 1981.

51. Sunday Times, December 2 1979.

52. ibid. April 6 1980 - HARPER of Joint Trawlers International.

53. Les Echos, November 12 1982.

54. European Court of Justice (ECJ) 359/74, Commission v France

55. Le Marin, July 13 1982.

56. Fishing News, July 10 and August 7 1981.

57. ibid. March 23 1984.

58. Le Marin, January 9 1981; cf Fishing News, January 16 1981. "La Grande Bretagne ne possède plus la flotte nécessaire...et fait ouvertement appel à des mercenaires ... espagnolols à qui l'on brade le pavillon brittanique pour tenter parvenir à les remplir."

59. Fishing News, December 18/25 1981.

60. ibid. February 12 1982.

61. Le Marin, October 10 1981.

CHAPTER NINE

1. MORDREL, L. (1972) op.cit. p 394.

2. LEIGH, M.(1983) op.cit. p 212.

3. Le Marin, January 27 1967 quoted in MORDREL, L. (1972) op.cit. p 777. "Trouvons-nous à Bruxelles et ailleurs des gens qui sont disposés à nous entendre? Au lieu d'armateurs et surtout de pêcheurs, nous ne recontrons le plus souvent comme interlocuteur que des répresentants des 'marchands' (i.e UNILEVER, FINDUS etc) lesquels ne s'intéressent guère au sort des producteurs...et pour cause!"

4. ibid. October 16 1981. "Si nous ne réagissons pas, dans quelques annèes la pêche française sera marginalisée ou intégrée par les circuits des multinationales."

5. cf AVERYT, W. (1965) 'Eurogroups, clientela and the European Community' in International Organisation, vol.29, no.4, pp 949-972.

6. cf BULMER, S. (1983) 'Domestic politics and European Community policy-making' in *Journal of Common Market Studies*, Vol.XXI, No.4, pp 349-363.

7. TUNSTALL, J. (1968) *Fish: An Antiquated Industry*, Fabian Tract 380, quoted in *Fishing Into The 80s*, p 19.

8. For a more detailed comparison, see *Fishing News*, November 5 1982: the final deal did not differ substantially from that then on offer.

9. quoted in HAYWARD, J.E.S. and BERKI, R.M. (eds) (1979) op.cit. p 18.

10. Interview with J. PLORMEL (30.9.83.).

11. BUTLIN, J.A. (1983) 'Do we really have a Common Fisheries Policy?' in *Fisheries Economics Newsletter*, SFIA, Edinburgh No.15, pp VII and VIII.

12. HOUSE OF COMMONS (1978) op.cit. Vol.II, p 28 - GILLETT.

13. STILES, G. (1976) 'The small maritime community and its resource management problems: a Newfoundland example" in JOHNSTON, D.M. (ed) op.cit. p 251.

14. *Le Marin*, May 8 1981 and March 26 1982.

15. RUGGIE, J.G. (1972) op.cit. p 888 ff.

Appendix II: List of abbreviations

AGEAM	Association de Gérance des Ecoles d'Apprentisage Maritime
ANOP	Association Nationale des Organisations de Producteurs
AWP	Autonomous Withdrawal Price
BFF	British Fishing Federation
BTF	British Trawlers' Federation
BUT	British United Trawlers
CAP	Common Agricultural Policy
CCPM	Comité Central des Pêches Maritimes
CEASM	Centre d'Etude et d'Action Sociales Maritimes
CES	Conseil Economique et Social
CFDT	Confédération Française Démocratique du Travail
CFP	Common Fisheries Policy
CGT	Confédération Générale du Travail
CNEXO	Centre National pour l'Exploitation des Océans
COCMM	Confédération des Organismes de Crédit Maritime Mutuel
DAFS	Department of Agriculture and Fisheries for Scotland
DATAR	Délégation à l'Aménagement du Territoire et à l'Action Régionale
DGMM	Direction Générale de la Marine Marchande
DOM-TOM	Départements d'Outre-Mer - Territoires d'Outre-Mer
DOT	Department of Trade
EAM	Ecole des Administrateurs Maritimes
EAPO	European Association of Producer Organisations (French initials AEOP)
ECJ	European Court of Justice
ECU	European Currency Unit
EEZ	Exclusive Economic Zone
ENA	Ecole Nationale d'Administration

ENIM	Etablissement National des Invalides de la Marine.
FEOGA	European Agricultural Guidance and Guarantee Fund (French initials)
FFSPM	Fédération Francaise des Syndicats Professionels Maritimes
FIOM	Fonds d'Intervention et d'Organisation des Marchés des Produits de la Pêche Maritime et de la Conchyliculture.
FOS	Fisheries Organisation Society (Ltd)
FROM	Fonds Régional d'Organisation du Marché du Poisson.
GATT	General Agreement on Tariffs and Trade.
GDP	Gross Domestic Product
HIB	Herring Industry Board
ICES	International Council for the Exploration of the Seas
MAFF	Ministry of Agriculture, Fisheries and Food
MSY	Maximum Sustainable Yield.
NEAFC	North East Atlantic Fisheries Commission
NFFO	National Federation of Fishermen's Organisations
OECD	Organisation for Economic Cooperation and Development
OJEC	Official Journal of the European Communities
OPEC	Organisation of Petroleum Exporting Countries
OREAM	Organisation d'Etudes d'Aménagement d'Aire Metropolitaine
ORSTOM	Office de la Recherche Scientifique et Technique d'Outre-Mer.
OWP	Official Withdrawal Price
PCF	Parti Communiste Français
PO	Producer Organisation
PROMA	Organisation de Producteurs de Pêche Artisanale du Morbihan et de la Loire Atlantique
PROMER	Comité National de Propagande pour la Consommation des Produits de la Mer
PSF	Parti Socialiste Français
RPR	Rassemblement pour la République

SFF	Scottish Fishermen's Federation
SFIA	Sea Fish Industry Authority
SFO	Scottish Fishermen's Organisation (Ltd)
SGMM	Secrétariat Général de la Marine Marchande
SNP	Scottish Nationalist Party
SNPL	Société Nouvelle des Pêches Lointaines
SODIPEB	Société pour le Développement des Industries de la Pêche en Bretagne
STF	Scottish Trawlers' Federation
TAC	Total Allowable Catch
TGWU	Transport and General Workers' Union
TOG	Trawler Officers' Guild
UAP	Union des Armateurs à la Pêche de France
UDF	Union pour la Démocratie Française
UKAFPO	United Kingdom Association of Fish Producer Organisations
UNCLOS	United Nations Conference on the Law of the Sea
WFA	White Fish Authority

The following abbreviations for measures are also used:

cwt	hundredweight
ECU	European Currency Unit
FF	French franc
kg	kilogram
MF	million French francs
t.	tonnes

Appendix III: List of interviews

The following people, all actively involved in the fisheries sector, were interviewed in the course of the study. In each case, the date and place of the interview are accompanied by their most relevant position at the time.

BRITAIN

Name	Position	Time and Place
D.AITCHISON	Chief Executive, SFF	14.4.80 - Edinburgh
S.ANDREWS	Information Officer, Humberside County Council	27.3.80.-Hull
H.BARRETT	Editor, Fishing News	6.2.80.-London
D.CAIRNS	Fisheries Officer,TGWU	26.3.80.-Hull
J.CORMACK	Fisheries Secretary,DAFS	15.4.80.- Edinburgh
J.DAVIES	Public Relations Adviser to the BFF	27.3.80.-Hull
J.KELSEY	Deputy Secretary,MAFF	26.2.80.-London
R.KEMP	Head of Industrial Development Unit, Hull City Council	26.3.80.-Hull
T.LIMAN	Research Assistant, Planning Department, Humberside County Council	27.3.80.-Hull
A.LOUGH	District Inspector of Fisheries, MAFF	26.3.80.-Hull
W.MASON	Deputy Secretary, MAFF	7.2.84.-Hull
N.McKELLAR	Chief Economist, Fishery Economics Research Unit, WFA	11.4.80.- Edinburgh
I.McSWEEN	Deputy Chief Executive, SFO	14.4.80.- Edinburgh
W.OAKESHOTT	Fisheries Economist, DAFS	15.4.80- Edinburgh
T.OLIVER	East Coast correspondent, Fishing News	22.9.83.-Hull
W.ROBB	Regional Officer, WFA	26.3.80.-Hull

FRANCE

M.BENOISH	Président, ANOP	1.12.83.-Paris
M.CLAIROUIN	Administrateur, FIOM	18.11.83.-Paris
M.DION	Secrétaire, Syndicat National des Thoniers Congélateurs	19.7.83-Concarneau
B.DUBREUIL	Président, CCPM	11.5.83.-Paris
J-P.GRANDIDIER	Sous-Directeur, Coopérative Maritime Etaploise	30.9.83.-Boulogne
C.GROUHEL	Sous-Directeur, PROMA	20.7.83.-Lorient
A.GRUENAIS	Délégué responsable des problemes maritimes internationaux et des pêches maritimes, Fédération Nationale des Syndicats Maritimes - CGT	15.4.83.-Paris
Y.GUILLEMONT	Secrétaire, Groupement des Armateurs à la Pêche Hauturière de Bretagne	19.7.83.-Concarneau
J.C.HENNEQUIN	Conseiller Technique, Secrétariat d'Etat à la Mer	28.10.83.-Paris
J.HURET	Vice-Président, CCPM et Président Directeur Général, Pêcheries de la Morinie	30.9.83.-Boulogne
M.LE BELLER	Secrétaire-Général, FIOM	23.7.82.-Paris
Y.L'HELGOUACH	Président, Comité Local des Pêches Maritimes de Concarneau, CGT.	19.7.83.-Concarneau
L.LEROUX	Président, Comité Local des Pêches Maritimes du Guilvinec	19.7.83.-Loctudy
G.MARCHAND	Conseiller Maritime, French Embassy	31.7.79.-London
J.PLORMEL	Directeur, FROM-Nord	30.9.83.-Boulogne
P.SOISSON	Secrétaire-Général, UAP	1.10.80.-Paris

Bibliography

The material used for this study is divided here into primary and secondary sources. The first category covers exclusively documents related to the fishing industry; the second category contains literature linked both to the issue area and to the more general themes discussed in this study.

Primary Sources

The main primary sources used were the general and specialized press on the fishing industry in the two countries, chiefly over the period from 1975 to the end of 1982. In Paris the press dossier 'Pêche' of the Fondation Nationale des Sciences Politiques was supplemented by consulting the back copies of Le Marin, the weekly newspaper of the industry and La Pêche Maritime, a monthly periodical. In London the weekly newspaper Fishing News complemented the EEC fisheries file of the Royal Institute of International Affairs.

Statistics were derived from a variety of annual publications: the HMSO series of Sea Fishery Statistical Tables, the Statistique de la Marine Marchande, the Eurostat material, in particular the Yearbook of Fisheries Statistics of the Statistical Office of the European Communities and the Bulletin Statistique of the International Council for the Exploration of the Seas (ICES).

A whole variety of documents produced by those working within the industry or by those investigating its plight was also used. They can be conveniently divided into sections covering material that originated in Britain, France or international institutions.

British material

BEHRENDT, M. (1980) Revision of the Common Fisheries Policy, paper delivered at the conference on Technology and Challenges of the World's New Fisheries Regime, 10 June 1980.

BRITISH UNITED TRAWLERS (1976) Proposals for a United Kingdom Fisheries Policy, Hull.

HMSO (1975) Food From Our Own Resources, Cmnd 6020.

HMSO (1982) Falkland Islands Economic Study 1982, Cmnd.8653 Chairman-The Rt.Hon.Lord Shackleton.

HOUSE OF COMMONS (1978) The Fishing Industry, Fifth Report of the Expenditure Committee (Trade and Industry Sub-Committee) Session 1977-78.

HOUSE OF LORDS (1980) EEC Fisheries Policy, 67th Report of the Select Committee on the European Communities, Session 1979-80.

HUMBERSIDE COUNTY COUNCIL (1976) The Humberside Fishing Industry, Report for HM Government, June 11 1976.

MACKINTOSH, J.P. The Impact of EEC Policies on British Fishing and Agriculture, European Movement (undated).

NFFO newsletters to members in 1979 and 1980.

PRESCOTT, J. (1978) Fishing into the 80s - A Discussion Document, prepared by group chaired by J. PRESCOTT and including D. Blackman, D. Cairns, J. Godfrey, N. Godman and J. Stoddart.

SFF undated newsletters to members.

SFIA (1983) The Administration of the Fishing Industry in France at Central and Local Level, Occasional Paper Series No.1, Fishery Economics Research Unit, Edinburgh, (prepared by I.Milligan).

WFA (1972) Industry Participation in the Administration of the Fishing Industry in France, No.72/3, Fishery Economics Research Unit, Edinburgh (prepared by A.D. Insull and D.A. Palfreman).

WFA (1977) Fisheries of the European Community, Fishery Economics Research Unit, Edinburgh.

WFA (1978) A Case for Aid for Restructuring the Catching Sector, Edinburgh, Fishery Economics Research Unit, (internal document).

WFA (1980) The Fishing Industry: Some Facts and Figures (internal document)

A selection of the Commons' debates on fishing during this period (recorded in HANSARD) was also read.

French material

CFDT - 'La voix des Gens de Mer' (union newspaper covering most of the period of this study). Also 'Rapport Général: Activités 1975-78'.

CCPM - Rapports d'Activités and Rapports sur la Production (an invaluable source for the years 1971 to 1983).

CCPM (1982) Statuts de l'Organisation Professionelle des Pêches Maritimes, May 1982.

CONSEIL ECONOMIQUE ET SOCIAL (1976) L'Avenir des Pêches Maritimes, presented by J. MARTRAY.

LA DOCUMENTATION FRANÇAISE (1970) 'L'industrie des pêches brittaniques' in 'Les activités maritimes du Royaume Uni, de Grande Bretagne et d'Irlande du Nord', Notes et Etudes Documentaires, No.3727, pp 37-47.

LA DOCUMENTATION FRANCAISE (1978) 'Les Etats et la Mer-le nationalisme maritime', Notes et Etudes Documentaires, Nos.4451-4452, (prepared by L.LUCCHINI and M.VOELCKEL)

LA DOCUMENTATION FRANÇAISE (1978) 'La crise des industries de la pêche' in Problèmes Economiques, No.1585, pp 15-19.

LA DOCUMENTATION FRANCAISE (1980) Rapport du Groupe de Travail: Mer et Littoral, Préparation du Huitième Plan, 1981-1985.

LA DOCUMENTATION FRANÇAISE (1980) 'VIIIe Plan: le dossier de la pêche française, in Regards sur l'Actualité, No.66, pp 23-28.

LA DOCUMENTATION FRANÇAISE (1980) 'Vers un marché commun de la pêche' in Regards sur l'Actualité, No.36, pp 29-32.

LA DOCUMENTATION FRANÇAISE (1982) 'Gérer La Mer', in Les Cahiers Français, October-December 1982, No.208.

LA DOCUMENTATION FRANÇAISE (1983) 'L'Europe bleue' in Regards sur l'Actualité, No.91, pp 27-40.

FIOM (1981) Note d'Information sur le FIOM.

FIOM Rapports Annuels (for the years 1976-1982).

MINISTERE DES TRANSPORTS, Voies - magazine of the ministry.

LE MONDE (1981) 'La Pêche', Dossiers et Documents, No. 80.

LE MONDE (1981) 'L'Election Présidentielle 26 avril-10 mai 1981', Supplément aux Dossiers et Documents du Monde, May 1981.

OREAM (1982) La Filière des Produits de la Mer dans l'Ouest (prepared by M. Bru and B. Simon for DATAR, the regional planning agency.)

SECRETARIAT D'ETAT A LA MER (1983) Investir à la Pêche (brochure on fishing aids).

SECRETARIAT D'ETAT A LA MER (1983) Les Pêches Maritimes et les cultures maritimes françaises en 1982 (tables and maps of French catches).

SGMM (1977) Les Pêches Maritimes Françaises en 1976.

SODIPEB (1981) Etat de la Flotte des Chalutiers Hauturiers du Sud-Bretagne au Ier Janvier 1981 et Evolution des Flotilles au Cours de la Décennie 1970-1980. (internal document).

UAP (1975) 'La crise' in Germes, No.19, pp 1-27, (review of the French fishing boat owners).

International material

COMMISSION DES COMMUNAUTES EUROPEENNES (1970) Formes de Coopération dans le Secteur de la Pêche, Informations internes sur l'agriculture, December 1970, Bruxelles, nos.68 and 69.

COMMISSION OF THE EUROPEAN COMMUNITIES (1976) Forms of Cooperation in the Fishing Industry, Information on Agriculture, April 1976. no.9, Brussels.

COMMISSION DES COMMUNAUTES EUROPEENNES (1979) Impact Regional de la Politique de la Pêche de la CEE - Bretagne, Informations internes sur la pêche, Bruxelles.

OECD (1980) Financial Support to the Fishing Industry, Paris, OECD

OECD (1980) Review of Fisheries in OECD Member Countries 1979, Paris, OECD. (In French, Examen des Pêcheries dans les Pays Membres).

A selection of debates from the European Parliament, to be found in the Annexes of the Official Journal of the European Communities (OJEC) was also read.

Secondary Sources

Books and articles

ALLEN, R. (1980) 'Fishing for a common policy' in Journal of Common Market Studies, Vol.XIX, No.2, pp 123-139.

ANDREWS, B. (1975) 'Social rules and the state as a social actor' in World Politics, Vol.XXVII, No.4, pp 521-540.

ANDREWS, W.G. and HOFFMANN, S. (eds) (1980) The Fifth Republic at Twenty, New York, State University of New York Press.

ARCHER, T.C. (1978) 'New departures in the North Sea' in Co-operation and Conflict, XIII, pp 1-19.

ARDAGH, J. (1977) The New France, Harmondsworth, Penguin.

AVERYT, W. (1975) 'Eurogroups, clientela and the European Community' in International Organisation, Vol.29, No.4, pp 949-972.

BACON, R, and ELTIS, W. (1976) Britain's Economic Problem: Too Few Producers?, London, Macmillan.

BARBER, J.P. (1976) Who makes British foreign policy? Open University Press.

BARNES, S.J. and SCASE, M. (1979) Political Action : Mass Participation in Five Western Democracies, Beverley Hills, California.

BECAM, M. (1976) 'Une region dont la population dépend étroitement des richesses de la mer: la Bretagne' in Revue Iranienne des Relations Internationales, Nos.5-6, Winter '75-'76.

BEER, S.H. (1965) British Politics in the Collectivist Age, New York, Knopf.

BENEWICK, R. and SMITH, T. (1972) Direct Action and Democratic Politics, London, George Allen and Unwin.

BERGER, S. (1981) Organising interaction in Western Europe, Cambridge, Cambridge University Press.

BIDET, J. (1974), 'Sur les rapports d'être de l'idéologie, les rapports sociaux dans le secteur de la pêche' in La Pensée No.174, March-April, pp 53-66.

BIRCH, A.J. (1964) Representative and Responsible Government, London, George Allen and Unwin.

BIRNBAUM, P. (1982) La Logique de l'Etat, Paris, Fayard.

BLOCH, J.P. and FOURNET, P. (1977) 'Le nouveau droit de la mer et ses incidences sur les pêches maritimes françaises' in Les Cahiers d'Outre-Mer, 30 (120), October-December, 1977 pp 326-347.

BOARDMAN, R. (1976) 'Ocean politics in Western Europe' in JOHNSTON, D.M. (ed.) Marine Policy and the Coastal Community, London, Croom Helm, pp 183-213.

BORNSTEIN, S. HELD, S. and KRIEGER, J. (eds) (1984) The State in Capitalist Europe, London, George Allen and Unwin.

BORTHWICK, R.L. and SPENCE, J.E. (eds) (1984) British Politics in Perspective, Leicester, Leicester University Press.

BOYER, A. (1967) Les Pêches Maritimes, Paris, Presses Universitaires de France (Que Sais-Je? No.199)

BRITTAN, S. (1975) 'The economic contradictions of democracy' in British Journal of Political Science, Vol.5, No.1, pp 129-159.

BROWN, B.E. (1969) 'The French experience of modernisation' in World Politics, Vol.XXI, No.3, pp 366-391.

BROWN, E.D. (1977) The Exclusive Economic Zone and the Law of the Sea, London, Taylor and Francis.

BULMER, S. (1983) 'Domestic politics and European Community policy-making' in Journal of Common Market Studies, Vol.XXI, No.4, pp 349-363.

BURDEAU, G. (1970) L'Etat, Paris, Seuil.

BURROWS, B. DENTON, G. and EDWARDS, G. (eds) (1977) Federal Solutions to European Issues, London, Macmillan.

BUTLIN, J.A. (1975) 'The political economy of international marine resources management' in Resources Policy, Vol.2, No.2, pp 128-135.

BUTLIN, J.A. (1983) 'Do we really have a Common Fisheries Policy?' in Fisheries Economics Newsletter, SFIA, Edinburgh, No.15, pp II-X.

BUZAN, B. (1978) A sea of troubles?: sources of dispute in the new ocean regime, London, International Institute for Strategic Studies (Adelphi papers No.143).

CANETTI, E. (1981) Crowds and Power, Harmondsworth, Penguin.

CARGILL, G. (1976) Blockade '75: the Story of the Fishermen's Blockade of the Ports, Glasgow, Molendinar Press.

CAWSON, A. (1978) 'Pluralism, corporatism and the role of the state' in Government and Opposition, Vol.13, No.2, pp 178-198.

CERNY, P.G. and SCHAIN, M.A. (eds) (1980) French Politics and Public Policy, London, Frances Pinter.

CHAPIN, Jean-Yves (1972) La Politique Commune Des Pêches, Thèse de Doctorat, Rennes.

CHOURAQUI, G. (1979) La Mer Confisquée, Paris, Seuil.

CHRISTY, F.T. Jr. (1977) 'Transition in the management and distribution of international fisheries' in International Organisation, Vol.31, No.2, pp 235-265.

CHURCHILL, R. SIMMONDS, K.R. and WELCH, J. (eds) (1973) New Directions in the Law of the Sea, New York, Oceana Publications.

CHURCHILL, R. (1977) 'The EEC fisheries policy - towards a revision' in Marine Policy, Vol.1, No.1, pp 26-36.

CHURCHILL, R. (1980) 'Revision of the EEC's Common Fisheries Policy,' - Part I, Marine Policy Vol.5, No.1, pp 3-37.

CHURCHILL, R. (1980) 'Revision of the EEC's Common Fisheries Policy,' - Part II, Marine Policy Vol.5, No.2, pp 95-111.

CIRIACY-WANTRUP, S.V. and BISHOP, R.C. (1975) '"Common Property" as a concept in natural resources policy' in Natural Resources Journal, Vol.15, No.4, pp 713-727.

CLUB SOCIALISTE DU LIVRE (1981) La Mer Retrouvée, Paris.

COHEN, S.S. (1977) Modern Capitalist Planning: The French Model, Berkeley, University of California Press.

COHEN, S.S. and GOUREVITCH P.A. (eds) (1982) France in the Troubled World Economy, London, Butterworth.

COOMBES, D. (1982) Representative Government and Economic Power, London, Heinemann.

COOPER, R.M. (1972) 'Economic interdependence and foreign policy in the seventies' in World Politics, Vol.24, No.2, pp 159-181.

CORLAY, J.P. (1983) 'La grève des marins-pêcheurs de l'été 1980 en Basse Normandie. Géographie d'un conflit social en milieu halieutique' in Façade Atlantique - Les Faits d'Occupation Conflictuelle du Littoral, Université de Nantes, Groupe de Recherche sur les Structures Economiques et les Rapports Sociaux, pp 196-234.

COULL, J.R. (1972) The Fisheries of Europe, London, G. Bell and Sons Ltd.

COUPER, A.D. (1978) 'Marine resources and environment' in Progress in Human Geography, Vol.2, No.2, pp 296-308.

COX, A. (ed) (1982) Politics, Policy and the European Recession, London, Macmillan.

COX, A. and HAYWARD, J. (1983) 'The inapplicability of the corporatist model in Britain and France' in International Political Science Review, Vol.4, No.2, pp 217-240.

CROUCH, C. (ed) (1979) State and Economy in Contemporary Capitalism, London, Croom Helm.

CUSHING, D. (1975) Fisheries Resources of the Sea and their Management, Oxford, Oxford University Press.

DADRE, V.C. (1980) Vers une Politique Commune de la Pêche dans la Communauté: Fondements Juridiques et Problemes d'Elaboration, Thèse pour le Doctorat du 3e cycle, Université de Bordeaux, Faculté de Droit et des Sciences Economiques.

DEBEAUVAIS, R. (1983) 'La place de l'artisanat dans la société française: l'exemple de la pêche maritime' in Dossier pour notre temps, Nos.20-21, May-August 1983. article 1, pp 1-4.

DRISCOLL, D.J. and McKELLAR, N. (1979) 'The changing regime of North Sea fisheries' in MASON, C.M. (ed) The Effective Management of Resources, London, Frances Pinter, pp 125-167.

DUHAMEL, G. and HUREAU, J.C. (1981) 'La situation de la pêche aux îles Kerguelen en 1981' in La Pêche Maritime, May 1981, pp 272-9.

DYSON, K. (1979) 'The ambiguous politics of Western Germany' in European Journal of Political Research, Vol.7, No.4, pp 375-96.

DYSON,K. (1980) The State Tradition in Western Europe, Oxford, Martin Robertson.

ECONOMIC AND SOCIAL COMMITTEE (1980) European Interest Groups and their Relationships with the Economic and Social Committee, Farnborough, Saxon House.

EISENHAMMER, J. (1983) The Parti Communiste Français And the Confédération Générale du Travail in Contemporary French Politics: A Study of Some Aspects of the Organizations And Their Relationship, D.Phil Thesis, Hilary Term 1983, Nuffield College, University of Oxford.

ESNOUF, B. (1976) La Crise de l'Industrie des Pêches Maritimes en 1975, CCPM, Ecole Nationale Supérieure Agronomique de Rennes, Mémoire de fin d'études.

FARNELL, J. and ELLES, J. (1984) In Search of a Common Fisheries Policy, Aldershot, Gower.

FAWCETT, J.E.S. (1977) 'So UNCLOS failed - or did it?" in The World Today, Vol.33 No.1 pp 28-34.

FELD, W.J. (1966) 'National economic interest groups and policy formation in the European Economic Community' in Political Science Quarterly, 81, pp 392-411.

FINER, S.E. (1974) 'Groups and political participation' in KIMBER, R. and RICHARDSON, J.J. (eds) Pressure Groups in Britain, London, Dent, pp 255-275.

FOURNET, P. (1977) 'La grande pêche morutière française' in La Pêche Maritime, May 1977, pp 271-277.

FOURNET. P. (1978) 'La crise des pêches maritimes brittaniques' in La Pêche Maritime, October 1978, pp 569-574.

FREESTONE, D. (1983) 'The Common Fisheries Policy'in LODGE, J. (ed) The European Community: Bibliographical Excursions, London, Frances Pinter, pp 107-116.

FREESTONE, D. with FLEISCH, A. (1983) 'The Common Fisheries Policy' in LODGE, J. (ed) Institutions and Policies of the European Community, London, Frances Pinter, pp 77-84.

GERSCHENKRON, A. (1962) Economic Backwardness in Historical Perspective: A book of Essays, Cambridge, Mass; Harvard University Press.

GILCHRIST, A. (1978) Cod Wars and How to Lose them, Edinburgh, Q.Press.

GILPIN, R. (1968) France in the Age of the Scientific State, Princeton, Princeton University Press.

GILPIN, R. (1976) US Power and the Multinational Corporation, London, Macmillan.

GILPIN, R. (1975) 'Three models of the future' in International Organisation, Vol.29, No.1, pp 37-60.

GRANT, W. (1981) 'The politics of the green pound' in Journal of Common Market Studies,Vol.XIX, No.4, pp 315-329.

GREEN, D. (1978) 'Individualism versus collectivism: economic choices in France' in West European Politics, Vol.1, No.3, pp 81-96.

HAAS, E.N. (1975) 'On systems and international regimes' in World Politics, Vol. XXVII, No.2, pp 147-174.

HABERMAS, J. (1976) Legitimation Crisis, Heinemann, London.

HARRISON, R.J. (1980) Pluralism and Corporatism, London, George Allen and Unwin.

HAYWARD, J.E.S. (1972) 'State intervention in France: the changing style of government-industry relations' in Political Studies, Vol.40, No.3, pp 287-298.

HAYWARD, J.E.S. (1974) 'National aptitudes for planning in Britain, France and Italy' in Government and Opposition, Vol.9, No.4, pp 397-410.

HAYWARD, J.E.S. and WATSON, M. (eds) (1975) Planning, Politics and Public Policy: the British, French and Italian Experience, Cambridge, Cambridge University Press.

HAYWARD, J.E.S. (1976) 'Institutional inertia and political impetus in France and Britain' in European Journal of Political Research, Vol.4, No.4, pp 341-359.

HAYWARD, J.E.S. (1978) 'Dissentient France: the counter political culture' in West European Politics, Vol.1, No.3, pp 53-67.

HAYWARD, J.E.S. and BERKI, R.M. (eds) (1979) State and Society in Contemporary Europe, Oxford, Martin Robertson.

HAYWARD, J.E.S. (1982) 'Mobilising private interests in the service of public ambitions: the salient element in the dual French policy style?' in RICHARDSON, J. (ed) Policy Styles in Western Europe, London, George Allen and Unwin, pp 111-140

HAYWARD, J.E.S. (1982) 'France: the strategic management of impending collective impoverishment' in COX, A. (ed) Politics, Policy and the European Recession, London, Macmillan.

HAYWARD, J.E.S. (1983) Governing France: The One and Indivisible Republic, London, Weidenfeld and Nicolson.

HECK, C.N. (1978) 'Collective arrangements for managing ocean fisheries' in International Organisation, Vol.29, No.3, pp 711-743.

HODGES, M. (ed) (1972) European Integration, Harmondsworth, Penguin.

HOFFMANN, S. (1963) In Search of France, Cambridge, Mass; Harvard University Press.

HOOD, C. (1973) 'British fishing and the Iceland saga' in Political Quarterly, Vol.44, No.3, pp 349-352.

HOOD, C. (1978) 'The politics of the biosphere: the dynamics of fishing policy' in ROSE, R. The Dynamics of Public Policy, London and Beverley Hills, Sage, pp 57-79.

HOOD, C. and WRIGHT, M. (eds) (1981) Big Government in Hard Times, Oxford, Martin Robertson.

HURET, J. (1976) 'La pêche dans les perspectives économiques de la France' in La Pêche Maritime, February 1976, pp 202-226.

IONESCU, G. (1975) Centripetal Politics, London, Hart-Davis MacGibbon.

JAEGGER, C. (1982) Artisanat et Capitalisme, l'Envers de la Roue de l'Histoire, Paris, Payot.

JANIS, M.W. (1976) Sea Power and the Law of the Sea, Lexington, Lexington Books.

JOHNSON, B. (1975) 'Technocrats and the management of international fisheries' in International Organisation, Vol.29, No.3, pp 745-770.

JOHNSON, B. and ZACHER, M.W. (eds) Canadian Foreign Policy and the Law of the Sea, Vancouver, University of British Columbia Press.

JOHNSTON, D.M. (ed) (1976) Marine Policy and the Coastal Community: The Impact of the Law of the Sea, London, Croom Helm.

JORDAN, G. and RICHARDSON, J. (1982) 'The British policy style or the logic of negotiation' in RICHARDSON, J. (ed) Policy Styles in Western Europe, London, George Allen and Unwin.

JORION, P. (1983) Les pêcheurs d'Houat, Paris, Collection Savoir, Hermann.

KATZENSTEIN, P.J. (1976) 'International relations and domestic structures: foreign economic policies of advanced industrial states' in International Organisation, Vol.30, No.1, pp 1-45.

KEOHANE, R.O. and NYE, J.S.(1977) Power and Interdependence, Boston, Little Brown.

KESSELMAN, M. (1970) 'Overinstitutionalisation and political constraint. The case of France' in Comparative Politics, Vol.3, No.1, pp 21-44.

KIMBER, R. and RICHARDSON, J.J. (eds) (1974) Pressure Groups in Britain, London, Dent.

KIMBER, R. and RICHARDSON, J.J. (eds) (1974) Campaigning for the Environment, London, Routledge and Kegan Paul.

KINDLEBERGER, C.P. (1973) The World in Depression, 1929-1939, London, Allen Lane.

KING, A. (1975) 'Overload: problems of governing in the 1970s' in Political Studies, Vol.XXXIII, Nos 2-3, pp 162-174.

KIRCHNER, E. and SCHWAIGER, K. (1981) The Role of Interest Groups in the European Community, Aldershot, Gower.

LAING, A. (1971) 'The Common Fisheries Policy of the Six' in Fish Industry Review, Vol.1, No.2, pp 8-18.

LAMBERT, J. (1974) 'The politics of fisheries in the Community' in The European Economic Community, A Post-Experience Course, The Open University, Block 3, pp 135-152.

LAVANANT, M.G. (1980) 'La Pêche' in Sciences et Avenir, No.405, November, 1980, pp 42-61.

LE BIHAN, (1977) Organisations de Producteurs des Pêches Maritimes en France et Droit Communautaire, CNEXO, Rapports Economiques et Juridiques, No.5.

LEIGH, M (1983) European Integration and the Common Fisheries Policy, Croom Helm, London.

LIEBER, R.J. (1974) 'Interest groups and political integration' in KIMBER, R and RICHARDSON, J.J. (eds) Pressure Groups in Britain, London, Dent, pp 27-56.

LODGE, J. (ed.) (1983) The European Community: Bibliographical Excursions, London, Frances Pinter.

LODGE, J.(ed) (1983) Institutions and Policies of the European Community, London, Frances Pinter.

LOWI, T. (1972) 'Four systems of policy, politics and choice' in Public Administration Review, July-August 1972, pp 298-310.

LYSENKO, V. (1983) A Crime Against the World: Memoirs of a Russian Sea Captain, London, Gollancz, (trans.M. Glenny).

MABIRE, J. (1975) Pêcheurs du Cotentin, Caen, Editions Heimdal.

MACRIDIS, R.C. (1961) 'Interest groups in comparative analysis' in Journal of Politics, Vol.23, No.1, pp 25-45.

MASON, C.M. (ed) (1979) The Effective Management of Resources, London, Frances Pinter.

MEYNAUD, J. (1962) Nouvelles Etudes sur les Groupes de Pression, Paris.

MEYNAUD, J. and SIDJANSKI, D. (1971) Les Groupes de Pression dans la CEE, Bruxelles, Institut de Sociologie de l'Université Libre de Bruxelles.

MIDDLEMAS, K. (1979) Politics in Industrial Society, London, Deutsch.

MILIBAND, R. (1973) The State in Capitalist Society, London, Quartet.

MORDREL, L. (1972) Les Institutions de la Pêche Maritime, Thèse pour le Doctorat en Droit, Paris II, (2 volumes).

MORSE, E.L. (1973) Foreign Policy and Interdependence in Gaullist France, Princeton, Princeton University Press.

OLIVER, T.A. (1982) The Management and Organisation of the Fish Catching Industry on the North East Coast of England, A Research Report from the Centre for Fisheries Studies at Hull College of Higher Education.

OLSON, M. (1965) The Logic of Collective Action, Cambridge, Harvard University Press.

PARRES, A. (1976) 'Les vraies ressources de la pêche' in Projet, 104, April 1976, pp 414-420.

PEYREFITTE, A. (1976) Le Mal Français, Paris, Plon.

PIPER, S. (1974) The North Ships: The Life of a Trawlerman, Newton Abbott, David and Charles.

PRESCOTT, J.V.R. (1975) The political geography of the oceans, London, David and Charles.

RECHER, J. (1977) Le Grand Métier, journal d'un capitaine de pêche à Fécamp, Paris, Plon.

REJAI, M. and ENLOE, C.H. (1969) 'Nation-states and state-nations' in International Studies Quarterly, Vol.13, No.2, pp 140-58.

RICHARDSON, J. (ed) (1982) Policy Styles in Western Europe, London, George Allen and Unwin.

RIDLEY, F.F. (1970) Revolutionary Syndicalism in France, Cambridge, Cambridge University Press.

RUGGIE, J.G. (1972) 'Collective goods and future international collaboration' in The American Political Science Review, Vol.66, No.3, pp 874-901.

RUGGIE, J.G. and HAAS, E.N. (1975) 'International responses to technology' in International Organisation, Vol.29, No.3, pp 557-584.

RUGGIE, J.G. (1982) 'International regimes, transactions and change: embedded liberalism in the postwar economic order' in International Organisation, Vol.36, No.2, pp 379-415.

SACKS, P.M. (1980) 'State structure and the assymetrical society' in Comparative Politics, Vol.12, No.3, pp 349-76.

SCASE, R. (1980) The State in Western Europe, London, Croom Helm.

SCHMIEGELOW, H. and M. (1975) 'The new mercantilism in international relations: the case of France's external monetary policy' in International Organisation, Vol.29, No.2, pp 367-92.

SCHMITTER, P.C. (1974) 'Still the century of corporatism?' in The Review of Politics, Vol.36, No.1, pp 85-131.

SCHWARTZENBERG, R.G. (1981) La Droite Absolue, Paris, Flammarion

SELF, P. and STORING, H.J. (1958) The State and the Farmer, London, George Allen and Unwin.

SHACKLETON, M. (1981) 'What made the French fishermen resort to open protest?' in Marine Policy, Vol.5, No.4, pp 340-343.

SHACKLETON, M. (1983) 'Fishing for a policy? The common fisheries policy of the Community' in WALLACE, H. WALLACE, W. and WEBB, C.(eds) Policy Making in the European Community, Chichester, John Wiley, pp 349-371.

SHERIFF, P. (1979) 'French administration: sanctified or demystified?' in West European Politics, Vol.2, No.2, pp 262-267.

SHONFIELD, A. (1965) Modern Capitalism, London, Oxford University Press

SIBTHORP, M.M. (ed) (1975) The North Sea: Challenge and Opportunity, Europa for the David Davies Memorial Institute of International Studies.

SIBTHOP, M.M. (ed) (1977) Oceanic Management: Conflicting Uses of the Celtic Sea and Other Western UK Waters, London, Europa for the David Davies Memorial Institute of International Studies.

SIDJANSKI, D. (1972) 'Pressure groups and the European Economic Community' in HODGES, M. (ed) European Integration, Harmondsworth, Penguin, pp 401-420.

SMITH, M. LITTLE, R. and SHACKLETON, M. (1981) Perspectives on World Politics, London, Croom Helm.

SOLLIE, F. et al.(1974) The Challenge of New Territories, Oslo, Universitatsforlaget.

SPERO, J.E. (1977) The Politics of International Economic Relations, London, George Allen and Unwin.

STILES, G. (1976) 'The small maritime community and its resource management problems:a Newfoundland example' in JOHNSTON, D.M. (ed) Marine Policy and the Coastal Community, London, Croom Helm, pp 235-254.

SULEIMAN, E.N. (1974) Politics, Power and Bureaucracy, Princeton, Princeton University Press.

SULEIMAN, E.N. (1977) 'The myth of technical expertise: selection, organisation, and leadership' in Comparative Politics, Vol.10, No.1, pp 137-158.

SULEIMAN, E.N. (1979) Elites in French Society, the Politics of Survival, Princeton, Princeton University Press.

TROEL, J.L. (1978) Contribution à l'Analyse de la Stratégie des Groupes Internationaux dans le Secteur des Produits de la Mer et de l'Aquaculture, INRA, (under the supervision of Denez L'Hostis) (2 volumes).

TUNSTALL, J. (1962) The Fishermen, London, MacGibbon and Kee.

TUNSTALL, J. (1968) Fish: An Antiquated Industry, Fabian Tract 380.

VERNON, R. (1974) Big Business and the State, Cambridge, Harvard University Press.

VOLLE, A. and WALLACE, W. (1977) 'How common a fisheries policy?' in The World Today, Vol.33, No.2, pp 62-72.

WALLACE, H. WALLACE, W. and WEBB, C. (eds) (1983) Policy Making in the European Community, Chichester, John Wiley.

WALLACE, W. (1983) 'Less than a federation, more than a regime: the Community as a political system' in WALLACE, H. WALLACE, W. and WEBB, C. (eds) Policy Making in the European Community, Chichester, John Wiley, pp 403-436.

WATT, D.C. (1977) The EEC and fishing: new venture into unknown seas' in Political Quarterly, Vol.48, No.3, pp 328-336.

WATT, D.C. (ed) (1980) The North Sea: A New International Regime? Greenwich Forum V, Records of an International Conference at the Royal Naval College, Greenwich, 2, 3 and 4 May, 1979.

WILSON, F. (1979) 'The revitalization of French parties' in Comparative Political Studies, Vol.XII, No.1, pp 82-103.

WILSON, F.L. (1983) 'Les groupes d'intérêt sous la cinquième république' in Revue Française de Science Politique, Vol.33, No.2, pp 220-253.

WRIGHT, V. (1978) The Government and Politics of France, London, Hutchinson.

YOUNG, E. and FRICKE, P. (1975) Sea Use Planning, Fabian Tract 437.

YOUNG, O.R. (1977) Resource Management at the International Level: The Case of the North Pacific, London, Frances Pinter,

YOUNG, O.R.(1978) 'Anarchy and social choice: reflections on the international polity' in World Politics, Vol.30, No.2, pp 241-263.

YOUNG, S. and LOWE, A.V. (1974) Intervention in the Mixed Economy, London, Croom Helm.

YOUNG, S.Z. (1973) Terms of Entry: Britain's Negotiation with the European Community, 1970-1972, London, Heinemann.

ZYSMAN, J. (1977) Political Strategies for Industrial Order: State, Market, and Industry in France, Berkeley, University of California Press.

ZYSMAN, J. (1977) 'The French state in the international economy' in International Organisation, Vol.31, No.4, pp 839-877.